T0253561

Technical Safety, Reliability and Resilience

Ivo Häring

Technical Safety, Reliability and Resilience

Methods and Processes

 Springer

Ivo Häring
Fraunhofer Institute for High-Speed Dynamics
Ernst-Mach-Institut, EMI
Freiburg im Breisgau
Baden-Württemberg, Germany

ISBN 978-981-33-4274-3   ISBN 978-981-33-4272-9   (eBook)
https://doi.org/10.1007/978-981-33-4272-9

This Springer imprint is published by the registered company Springer Nature Singapore Pte Ltd.
The registered company address is: 152 Beach Road, #21-01/04 Gateway East, Singapore 189721, Singapore

The preparation of this textbook has been supported in parts by the Federal Ministry of Education and Research (BMBF) within the Federal Government—German Federal State competition "Advancement through Education: Open Universities."

Acknowledgements

Because of the broad scope on risk control and resilience enhancement of (socio) technical systems, system modeling, analytical system analysis, functional safety, and development of reliable and safe systems that show resilience post disruptions, there is a broad range of persons and bodies to be thanked.

Parts of the contents have been presented within study courses to numerous students generating valuable feedback. This includes continuous academic training (certificate of advanced study) as developed by Fraunhofer Ernst-Mach-Institut (EMI), Freiburg, in collaboration with the University of Freiburg and the Fraunhofer Academy Munich, and numerous classical courses held at the University of Freiburg, at the Baden-Wuerttemberg Cooperative State University (DHBW), Lörrach, and at the Hochschule Furtwangen University of Applied Science (HFU) in the areas of technical safety, reliability, and functional safety and resilience engineering.

Thanks also go to the following bachelor, diploma, master, or Ph.D. students at Fraunhofer EMI who contributed to applied research projects, in particular, in the methodology tailoring and development, as referenced throughout the textbook: Ranganatha (2020), Sudheeran (2020), Satsrisakul (2018), Jain (2018), Michael (2014), Thielsch (2012), Meier (2011), Ringwald (2010), Voss (2011), Brocke (2010), Ettwein (2010), Schubert (2009), Barlow (2009), Wolf (2009), Neumann-Heyme (2008), Glose (2008), Larisch (2008), Hänle (2007), and Schmidt (2007).

The text implemented valuable feedback from Scientists and Heads of Research Groups of the Department of Security Technologies and Protective Structures at Fraunhofer EMI, including A. Thein, J. Kaufman, B. Kanat, J. Scheidereiter, G. Hormann, S. Ebenhöch, U. Siebold, J. Finger, M. Fehling-Kaschek, and K. Fischer.

The generation as well as the iteration of the text was very much supported by the scientific lector work of Sina Rathjen who compiled first versions of the text based on outlines by the author and content materials as provided in selected sources that were generated during research project works. Without her tremendous effort this work would not have been possible. Thanks also go to the librarian staff, the media, and the publishing team at the institute: J. Helbling, B. Soergel, U. Galena, B. Bindnagel, and J. Holz. The text owes much to the thorough support of the team at Springer Publishing House.

The author thanks the International Electrotechnical Commission (IEC) for providing permission to reproduce Information from its International Standards. All such extracts are copyright of IEC, Geneva, Switzerland. All rights reserved. Further information on the IEC is available from www.iec.ch. IEC has no responsibility for the placement and context in which the extracts and contents are reproduced by the author, nor is IEC in any way responsible for the other content or accuracy therein.

The Object Management Group (OMG) is thanked for the license to use Unified Modeling Language (UML) and Systems Modeling Language (SysML) sample diagrams and symbols.

Last but not least, I want to thank my family for all their patience in case of text and lecture preparation work and related absences: B. H., A. A. M. H., L. A. Y. H., K. J. A. H., J. U. B. H., and A. A. B. H.

References

Barlow, L D (2009): Unterstützung Sicherheitsnachweis durch Schaltungssimulation, Support of safety assessment using electronic circuit simulation. Diploma Thesis, Naturwissenschaftlich-technische Akademie Isney (Fachhochschule und Berufskollegs), Fraunhofer EMI, Efringen-Kirchen.

Brocke, Michael vor der (2010): Grundlagen für Sicherheitsanalyse von embedded Software zur Umsetzung des Auswertealgorithmus eines Radarabstandssensors, Fundamentals for the safety assessment of embedded software implementing an algorithm for radar distance measurement determination. Bachelor Thesis. Hochschule Coburg; Fraunhofer EMI Efringen-Kirchen.

Ettwein, Peter (2010): Schädigungsmodelle für stumpfe Verletzungen, Damage models for blunt injuries. Bechelor Thesis, Hochschule für Technik (HFT) Stuttgart, Fraunhofer EMI, Efringen-Kichen.

Glose, D (2008): Zuverlässigkeitsvorhersage für elektronische Komponenten unter mechanischer Belastung, Reliability prediction of electronic components for mechanical stress environment. Hochschule Kempten, University of Applied Sciences; Fraunhofer EMI, Efringen-Kirchen.

Hänle, A. (2007): Modellierung und Spezifikation von Anforderungen eines sicherheitskritischen Systems mit UML, Modeling and Specification of Requirements of a safety critical System with UML. Hochschule Konstanz für Technik, Wirtschaft und Gestaltung (HTWG), University of Applied Sciences; Fraunhofer EMI, Efringen-Kirchen. Computer Science.

Jain, Kumar Aishvarya (2018): Indoor ultrasound location system simulation for assessment of disruptive events and resilience improvement options. Master Thesis. FAU Erlangen-Nürnberg, Fraunhofer EMI.

Larisch, M. (2008): Unterstützung des Nachweises funktionaler Sicherheit nach IEC 61508 durch SysML, Support of the functional Safety assessment according to EC 61508 with SysML. Diploma Thesis. Hochschule Konstanz für Technik, Wirtschaft und Gestaltung (HTWG), University of Applied Sciences; Fraunhofer EMI, Efringen-Kirchen.

Meier, Stefan (2011): Entwicklung eines eingebetteten Systems für eine Sensorauswertung gemäß IEC 61508. Bachelor Thesis. Duale Hochschule Baden-Württemberg, Lörrach. Fraunhofer EMI.

Michael, Rajesh (2014): Electrical characterization of electronic components during high acceleration: analytical and coupled continuum models for characterization of loading and explaining experiments. Master Thesis. Jena, Ernst-Abbe-Fachhochschule, Fachbereich Sci-Tec: Präzision, Optik, Materialien. Fraunhofer EMI.

Neumann-Heyme, Hieram (2008): Numerische Simulation zum Versagen elektronischer Komponenten bei mechanischer Belastung, Numerical simulation of the failure of electronic components under mechanical stress. Diploma Thesis. Technical University (TU) Dresden; Fraunhofer EMI, Efringen-Kirchen.

Ranganatha, Madhusudhan (2020): Robust signal coding for ultrasound indoor localization. Master Thesis, SRH Hochschule Heidelberg, Fraunhofer EMI.

Ringwald, Michael (2010): Entwicklung der Sensorauswerteschaltung für eine Sicherungseinrichtung nach IEC 61508. Bachelor Thesis. Duale Hochschule Baden-Württemberg, Lörrach. Fraunhofer EMI.

Schmidt, A. (2007): Analyse zur Softwaresicherheit eines Embedded Systems mit IEC 61508, Analysis to assess the software safety of an Embedded System using IEC 61508. Diploma Thesis. Hochschule Furtwangen University (HFU), Fraunhofer EMI.

Schubert, Florian (2009): Maschinenbautechnische Analyse generischer mechanischer Teilsysteme bei hoher mechanischer Belastung, Machine engineering construction analysis of generic mechanical subsystems under high mechanical stress. Diploma Thesis, Rheinische Fachhochschule (RFH) Köln, Fraunhofer EMI, Efringen-Kirchen.'

Thielsch, P. (2012). Risikoanalysemethoden zur Festlegung der Gesamtssicherheitsanforderungen im Sinn der "IEC 61508 (Ed. 2)". Bachelor, Hochschule Furtwangen.

Voss, Martin (2011): Methodik zur Bestimmung des Verhaltens von elektronischen Komponenten unter hoher Beschleunigung, Method to determine the behaviour of electronic components under high acceleration. Master Thesis. University of Freiburg, IMTEK; Fraunhofer EMI, Efringen-Kirchen, Freiburg.

Wolf, Stefan (2009): Verhalten elektronischer Bauteile bei hochdynamischer Beschleunigung, Behaviour of electronic components under high-dynamic acceleration. Master Thesis, TU Chemnitz, Fraunhofer EMI, Efringen-Kirchen.

About This Book

Developers of safety-relevant and safety-critical technical systems, high reliability systems, or dependable systems are challenged by an ever-increasing list of requirements regarding the development process, methods, techniques, and measures to be employed. Also, safety and security standards, legislation and jurisdiction, and post-accident events point to the need of well-informed development processes and method selections, tailoring, and adaption.

The book introduces some of the most relevant processes and classical methods for developing reliable and safety-critical systems. In addition, it shows how the methods are expected to support achieving systems that are resilient post disruptions, i.e., efficiently coping with system (service) function loss post major damage events or faults. This goes beyond classical risk control approaches.

Selection criteria for methods include general applicability, relevancy for practical applications, practical feasibility, and academic acceptance. A focus is to go somewhat beyond standard development life cycles, safety development live cycles, and risk assessment and control schemes by inventorying and applying new perspectives of (technical) resilience engineering approaches, within the safety 2.0 discussion or resilience analysis and management approaches. It is indicated how to select appropriate processes and methods for the implementation of such concepts.

A strong focus is on analytical inductive system analysis methods, e.g., variations of hazard list (HL), hazard analyses (HA, SSHA, SHA, O&SHA), and failure modes and effects analysis (FMEA, FMEDA, FMECA). Double failure matrix (DFM), event analysis (EA), and hazard log are shortly introduced. A main focus is also on deductive analytical methods, in particular, fault tree analysis (FTA, TDFTA). The tabular and graphical methods are motivated. Prerequisites, limitations, and tailoring options are given. Evaluation options are described in detail to foster the practical impact of the methods.

Semi-formal graphical diagrams of the systems using unified and systems modeling language (UML, SysML) are presented as methods to document and model systems under design or redesign, sufficiently to apply system analysis methods. In particular, it is shown how to model requirements of various kinds using SysML.

Typical reliability prediction approaches are summarized. They are applied to generate parameters most relevant for functional safety developments, including safety integrity level (SIL), safe failure fraction (SFF), diagnostic coverage (DC), and dangerous undetected (DU) and detected (DD) failures.

Also, some basic methods are introduced for securing the integrity of data within electronic embedded systems, in particular, data stored with limited resources. It is shown how to assess the efficiency of the methods in operational hardware contexts.

A further focus is on the description of the functional safety life cycle according to IEC 61508, as well as the selection, combination, and joint tailoring of methods for efficiently fulfilling functional safety requirements, in particular, of all tabular methods introduced in detail.

Application domains of the processes and methods include automotive systems for autonomous driving functions, process, and production industry; industry automation and control systems; critical infrastructure systems; renewable energy generation, storage, and distribution systems; and energy and water smart grids.

The book addresses developers and students up to Ph.D. level who want to gain a fast overview and insight on practically feasible processes and methods for designing and developing safe, reliable, and resilient systems, in particular, in the domain of (more) sustainable systems. Some of the chapters summarize selected research project results and should also be of interest for academia, in particular, when relating processes and methods to novel safety and resilience concepts for technical systems.

Contents

Abbreviations

AADL	Architecture Analysis and Design Language
AFNOR	Association Française de Normalisation
ANSI	American National Standards
BB	Birnbaum
BCD	Binary Coded Decimal
BFA	Bouncing Failure Analysis
BOM	Base Object Model
BSI	British Standard Institute
C	Consequences
D	Column of the FMEA for relation to legal requirements
DC	Diagnostic coverage
DFM	Double Failure Matrix
DFTA	Dynamic Fault Tree Analysis
DIN	Deutsches Institut für Normung
DoD	Department of Defense
e.g.	For example
E/E/PE	Electrical/Electronic/Programmable Electronic
EEPROM	Electrically Erasable Programmable Read-Only Memory
ECT	Estimated number of computing operations (estimated computing time)
EMM	European Media Monitor
EN	European Norm
EPS	Emergency Power System
ES	Exposed Site
ESQRA-GE	German Explosive Safety Quantitative Risk Analysis
ETA	Event Tree Analysis
EUC	Equipment Under Control
EZRT	Entwicklungszentrum für Röntgentechnik
FEM	Finite Element Method
FHA	Fault Hazard Analysis
FIT	Failures in Time
FMEA	Failure Modes and Effects Analysis

FMECA	Failure Modes and Criticality Analysis
FMEDA	Failure Modes, Effects and Diagnostic Coverage Analysis
F-N-curve	Frequency-Number-curve
FSQRA	Fuze Safety Quantitative Risk Analysis
FTA	Fault Tree Analysis
FV	Fussell–Vesely
GTD	Global Terrorism Database
HL	Hazard Log
i.e.	That is
IBD	Internal Block Diagram
IEC	International Electrotechnical Commission
IED	Improvised Explosive Device
IEV	International Electrotechnical Vocabulary
INS	Immediate, Necessary and Sufficient
ISO	International Organization for Standardization
JRC	Joint Research Center
LPG	Liquefied Petroleum Gas
MIPT	Memorial Institute for the Prevention of Terrorism
MOB	Multiple Occurring Branch
MOE	Multiple Occurring Event
MooN	M-out-of-N redundancy, e.g. 2oo3
MTBF	Mean Time Between Failures
MTTF	Mean in time to the next failure
NASA	National Aeronautics and Space Administration
NEQ	Net Explosive Quantity
O&SHA	Operating and Support Hazard Analysis
ON	Österreichisches Normeninstitut
OPD	Object Process Methodology
OPL	Object Process Language
OPM	Object Process Methodology
P	Probability
PC	Parts Count
PD	Probability of Default
PES	Potential Explosive Site
PFD	Probability of failure on demand
PFH	Probability of failure per hour
PHA	Preliminary Hazard Analysis
PHL	Preliminary Hazard List
PSC	Preliminary, secondary and command
R	Risk
RADC	Rome Air Development Center
RAF	Red Army Faction
RAW	Risk Achievement Worth
Rem	Roentgen equivalent in man
RF	Risk Factor

RPN	Risk Priority Number
RPZ	Risikoprioritätszahl
rrse	Root relative square error
RRW	Risk Reduction Worth
SAE	Society of Automotive Engineers
SD	Sequence Diagram
SDOF	Single Degree of Freedom
SFF	Safe Failure Fraction
SHA	System Hazard Analysis
SIL	Safety Integrity Level
SISO	Simulation Interoperability Standards Organization
SMA	Simple Moving Average
SMD	State Machine Diagram
SNV	Schweizerische Normenvereinigung
SQL	Structured Query Language
SRS	Software Requirements Specification
SSHA	Subsystem Hazard Analysis
SS-SC	State of System–State of Component
START	National Consortium for the Study of Terrorism and Responses to Terrorism
Sv	Sievert
SysML	Systems Modeling Language
TED	Terror Event Database
TEDAS	Terror Event Database Analysis Software
TM	Team Member
TNT	Trinitrotoluene
TS-FTA	Takagi–Sugeno Fault Tree Analysis
UK	United Kingdom
UML	Unified Modeling Language
UNI	Ente Nazionale Italiano di Unification
V	Verification and validation
VBIED	Vehicle-Borne Improvised Explosive Device
VDA	Verband der Automobilindustrie
VHDL	Very High-Speed Integrated Circuit Hardware Description Language

Mathematical Notations

π	Adjustment factor
$\wedge$	And
ω	Angular frequency
$\bar{x}$	Arithmetic mean
$\rho(k)$	Autocorrelation function
$\hat{\gamma}(k)$	Autocovariance

ρ_k	Autocovariance coefficient
$|A|$	Cardinality of A
$\overline{A}$	Complement of A
Ω	Complete space
$P(B|A)$	Conditional probability
C	Consequences
r_{xy}	Correlation coefficient
R_{crit}	Critical risk quantity
D	Detection rating
$\emptyset$	Empty set
$\cup_{ex}$	Exclusive union
$E(t)$	Expected value
$f(t)$	Failure density
$F(t)$	Failure probability
$\lambda(t)$	Failure rate
F	Feasibility rating
f	Frequency
I	Impulse
$\cap$	Intersection
$\tilde{x}$	Median
$\mathbb{N}_0$	Non-negative integers
—	Not
#	Number of
O	Occurrence rating
$\vee$	Or
T	Period
$\mathbb{N}$	Positive integers
P, p	Pressure
P	Probability
Pr	Probit
Rb	Reliability
R	Risk
S	Severity rating
α	Smoothing constant
$\Delta\Omega$	Solid angle surface element
s_x	Standard derivation
$\subseteq$	Subset
$\cup$	Union
s_x^2	Variance
$\backslash$	Without

List of Figures

List of Tables

Chapter 1
Introduction and Objectives

1.1 Overview

The introductory chapter gathers several narratives why systematic processes, techniques, and measures, in particular, classical and modified analytical system analysis methods, are key for the ideation, concept and design, development, verification and validation, implementation and testing of reliable, safe and resilient systems.

It provides disciplinary structuring and ordering principles employed by the text, main dependencies of the chapters, and features of the text to increase readability and relevancy for practitioners and experts. Chapters are highlighted that can be understood as stand-alone introductions to key system analysis methods, processes, and method combination options.

More recent applied research backgrounds are provided in the domains of automotive engineering of electrical vehicles, indoor localization systems, and sustainable energy supply of urban quarters. It shows that the book provides selected analysis methods and processes sufficient for achieving reliable, safe, and to lesser extent also resilient systems by resorting to classical system analysis methods including their tailoring and expected extensions.

In summary, the chapter gathers introductory material ranging from a general motivation (Sect. 1.2), disciplinary structuring and ordering principles employed by the textbook (Sect. 1.3), main features of the text (Sect. 1.4) to research background, and research project results used (Sect. 1.5).

I. Häring, *Technical Safety, Reliability and Resilience*,
https://doi.org/10.1007/978-981-33-4272-9_1

1.2 Safe, Secure, and Resilient Technical Sustainable Systems

When considering technical properties of (socio) technical systems and devices, the list of requirements is ever increasing, by now including.

- reliability,
- availability,
- low carbon-dioxide footprint, i.e., sustainability regarding the environmental effects,
- maintainability, and
- reusability or recyclability.

In addition, modern systems are increasingly connected, intelligent, flexible, used over a long life time when compared with the development time, are increasingly adaptive, reconfigurable, and autonomous. Hence, extending the classical risk landscape potential risk events that need to be covered range from expected to unexpected or even unexampled (back swans, unknown unknowns), from accidental via intentional (e.g., sabotage) to malicious (e.g., terroristic), external to internal. The risk events concern safety of systems, physical and IT security, range from minor to larger, unfold in short up long time scales, are minor up to disruptive, and catastrophic.

As the threat landscape increases with respect to failure modes and scales of failures, the further classical requirement that (socio) technical systems adequately provide

- control of risks, i.e., safety and security,

  has to be amended and extended by providing

- resilience with respect to disruptive or (potential) damage events.

Resilience can be generated by engineering for resilience, e.g., resilience by design, by development approaches followed, by using appropriate techniques and measures during development, and appropriate maintenance of systems during operation that support resilient behavior.

Another motivation of the textbook is to enable the reader to flexibly use, tailor and extend existing and practically established methods (best practices) of technical safety in novel contexts, and for pursuing novel system analysis aims. To this end, the assumptions, boundary conditions, and restrictions of the methods and processes are described. A particular focus is on how to adopt the methods for supporting the development of resilient systems.

A further motivation is to support the best practice of processes and methods established for the development of safety-relevant or safety–critical systems. Examples include the automotive domain, e.g., autonomous driving functions, production, process, and logistics industry, e.g., industrial automation and control systems, and critical infrastructure systems (live lines), e.g., supply grids for electricity, gas, telecommunication, and the banking system.

1.3 Structure of Text and Chapter Contents Overview

The textbook contains the following chapters:

- With motivations of the use of the proposed methods within classical risk control and resilience generation, definitions on basic safety and risk terms of technical system, and overviews on system analysis methods, see Chaps. 2, 3 and 5;
- Describing analyses and development processes, in particular, development model properties and representative models mainly for software and the safety life cycle of the generic functional safety approach, see Chaps. 10 and 11;
- On system analysis methods, in particular, system modeling for analytical system analyses and the families of hazard analyses (HAs), failure mode and effects analyses (FMEAs), and fault tree analyses (FMEA), see Chaps. 4, 6, 7 and 8;
- On requirements and semi-formal graphical system modeling methods, see Chaps. 12, 13 and 14, covering requirements types, the unified modeling language (UML), and the systems modeling language (SysML).
- On selected techniques and measures for system development and their combination, see Chaps. 9, 15 and 16.

The chapter sequence starts with an (prospective) overview on the usability of mainly classical system analysis methods for resilience engineering (see Chap. 2). Then the text introduces basic and more advanced system safety and analysis definitions and concepts (see Chaps. 3 and 4) sufficient for the classical system analysis approaches, which are introduced and compared in an overview (see Chap. 5). Next, fault tree analysis (FTA), failure modes and effects analysis (FMEA), and hazard analysis (HAs) methods including hazard log are introduced in detail (see Chaps. 6, 7 and 8, respectively).

Moving from single methods to development and assessment processes, first system development process types are introduced (see Chap. 10) to prepare for the discussion of the Safety Life Cycle approach of IEC 61,508 Ed. 2 (IEC 61,508) (see Chap. 11).

As a key input for FMEA- and FTA-like methods, Chap. 12 presents reliability prediction approaches and related standards, which can be used to determine failure rates and probabilities. Further sample methods that can be used together with classical system analysis methods to ensure sufficient documentation of system understanding are the semi-formal graphical unified and systems modeling languages (UML and SysML) (see Chaps. 13 and 14).

Chapter 15 discusses how to combine FMEA and FTA in a practical application in the automotive domain. These sample methods can be understood as specific input for safety–critical or safety-relevant software and hardware developments. Finally, examples of methods for detecting and countering flaws of data integrity, e.g., bit flips, are presented in a didactic fashion (see Chap. 16).

It is noted that rather challenging chapters like on the applicability of the classical methods to generate resilient technical systems with expect to damaging events (see Chaps. 2) and 6 on fault tree analysis (FTA) are at rather early position in the text. The reason is that the former can serve as a statement of the potential of the methods and the later chapter as a formally well-based starting point to more elementary methods. In particular, the advantages but also disadvantages of FMEA with respect to FTA can be discussed on safe ground.

From a functional safety perspective according to the generic standard IEC 61,508, the generic overall safety generation process of functional safety (safety life cycle) is presented, which includes a nested software and hardware development processes (V-Models for SW and HW). For the determination of requirements of safety functions, selected methods are presented, in particular, hazard analysis. For system analysis of electronic/electronic programmable/and electric (eepe)-safety-related systems, system analysis methods like failure modes effects and diagnostic analysis (FMEDA) as well as further techniques and measures are presented including reliability prediction, FTA, error detection and correction, and semi-formal graphical system modeling. The overall terminology is strongly aligned with IEC 61,508.

In summary, independent of the functional safety concept a set of complementary and most often and most successfully used methods for reliable and safe system development methods is presented. In addition, for each method it is indicated how it can be used to generate resiliency of technical systems. Furthermore, key methods found important for safe and secure system development like semi-formal graphical modeling, data integrity assessment, and reliability prediction are presented.

1.4 Main Features of the Text

Each chapter starts with a section containing a broad introduction and critical overview on the contents covered. Then content sections follow. The ambition is to provide graphics for the illustration of key ideas as well as worked out examples. Each chapter concludes with questions and answers. Each chapter can, in principle, be read independent of the other chapters.

The following chapters have stronger links and should be read as sequence. Chaps. 3 and 4 both introduce system analysis definitions and key concepts for analytical system modeling and analysis. They should be read before any of the chapters that cover analysis methods, namely Chaps. 5 to 8. Chapter 10 should be read before Chap. 11 since the former provides system development processes and the latter the functional safety life cycle. Chapter 13 should be read before Chap. 14 since both cover semi-formal graphical system modeling approaches for the application case of safety development. All other chapters can be read stand-alone.

Chapter 2 can be read as an overview on application options of the methods for developing resilient systems. It can also be read after any of the chapters on analysis methods, techniques and measures for system development, Chaps. 6, 7, 8, 9, 13, 14 or 16, is finished to start the tailoring and modification of a selected method for supporting the development of resilient technical systems.

1.5 Sample Background Research Projects

The subsequent sections mention selected applied research projects that used methods as presented within the textbook to show representative application examples and domains.

1.5.1 Functional Safety of Heating and Cooling Systems in Electrical Vehicles

The German–French research project AllFraTech (German–French alliance for innovative mobility solutions) subproject InnoTherMS (innovative predictive high-efficient thermal management systems) deals with the functional safety of novel heating and cooling systems for small electric transport vehicles (InnoTherMS 2020a, b).

Novel risks are investigated that are associated with the use of waste heat of the battery, inverter, and breaking systems for heating the driver's cabin. The aim is to identify essential requirements for future developments and to find adequate overall design architectures in this regard, especially if safety functions (e.g., battery cooling, heat dissipating power electronics) and comfort functions (e.g., interior cabin heating) overlap.

The use of waste heat potentially collides with the safety–critical cooling functions for the battery. Of interest are those risks that are considered critical because, if no constructive improvements can be found, they require mitigating safety functions. The project identifies risks to define requirements for safety functions, e.g., with the hazard analysis on system level, see Chap. 8. It conducts initial analyses on high abstraction level regarding the overall safety architecture, e.g., using fault tree analysis (FTA), see Chap. 6. Furthermore, for selected safety designs, the reliability of the safety functions and their self-diagnostic level can be estimated using failure modes effects, diagnostics and criticality analysis (FMEDA), see Chap. 7.

1.5.2 Resilience Engineering of Multi-Modal Indoor Localization System

The pilot solution demonstrator project on multi-modal efficient and resilient localization for intralogistics, production, and autonomous systems (MERLIN 2020) first applies a systematic approach to analytical resilience analysis of technical systems as described (Häring et al. 2017a, b) (Resilienzmaße 2018). In addition to indoor localization via ultrasound (Bordoy et al. 2020; Ens et al. 2015), the technologies Bluetooth, ultra-wide band (UWB), radio-frequency identification (RFID), and optical methods are assessed. The aim is reliable localization taking application environments and potential disruptions into account.

Due to the generic challenge of best combination of technologies for better risk control and resilience, preliminary hazard lists are a reasonable starting point to identify potential hazards of sufficient precise localization. For identified potential design options, this application also asks for extensions of (preliminary) hazard analysis methods for each localization technology and given operational environments, see Chap. 8. For instance, the hazard (disruption) acoustic (ultrasound) noise can be inductively assessed regarding its relevancy for a subset of localization technologies to determine whether it is robust against such an expectable disruption, e.g., in production shop floors. Other sample disruptions include barriers preventing line-of-sight communication and coverage of localization tags.

The most promising set of technologies can be assessed to exclude single-point function failures using failure modes and effects criticality analysis (FMECA), see Chap. 7, where the criticality level is defined in terms of a certain loss of localization accuracy as well as by using FTA, see Chap. 6.

1.5.3 Reliability and Resilience for Local Power Supply Grids

Within a project on the development of an organic-computing-based approach to provide and improve the resilience of ICT (OCTIKT 2020), local software agents are developed for electrical power distribution grids for urban quarters or small regions. Organic computing approaches are further developed to allow for independent and "plug-in-and-work" grid elements such as electrical transformers, electrical energy storage systems, prosumer households, or electrical vehicles.

The project is an example where the formulation of a framework and related methods for improving resilience is challenging (Tomforde et al. 2019; Häring and Gelhausen 2018), which is supported when resorting to the generic system development options as described in Chap. 10 as well as the functional safety life cycle overview of Chap. 11. Also, the identification of relevant methods for the process steps is of strong interest, see Chap. 15 on the selection and interfacing of methods.

This is also an example where the understanding of requirements, see Chap. 12, and their modeling are important, in particular, by resorting to semi-formal graphical methods such as unified modeling language, see Chap. 13, and systems modeling language, see Chap. 14.

References

Bordoy, Joan; Schott, Dominik Jan; Xie, Jizhou; Bannoura, Amir; Klein, Philip; Striet, Ludwig et al. (2020): Acoustic Indoor Localization Augmentation by Self-Calibration and Machine Learning. In *Sensors (Basel, Switzerland)* 20 (4). https://doi.org/10.3390/s20041177.

Ens, Alexander; Hoeflinger, Fabian; Wendeberg, Johannes; Hoppe, Joachim; Zhang, Rui; Bannoura, Amir et al. (2015): Acoustic Self-Calibrating System for Indoor Smart Phone Tracking. ASSIST. In *International Journal of Navigation and Observation* 2015. https://doi.org/10.1155/2015/694695.

Häring, Ivo; Sansavini, Giovanni; Bellini, Emanuele; Martyn, Nick; Kovalenko, Tatyana; Kitsak, Maksim et al. (2017a): Towards a generic resilience management, quantification and development process: general definitions, requirements, methods, techniques and measures, and case studies. In Igor Linkov, José Manuel Palma-Oliveira (Eds.): Resilience and risk. Methods and application in environment, cyber and social domains. NATO Advanced Research Workshop on Resilience-Based Approaches to Critical Infrastructure Safeguarding. Dordrecht: Springer (NATO science for peace and security series. Series C, Environmental security), pp. 21–80. Available online at https://www.springer.com/de/book/9789402411225.

Häring, Ivo; Scheidereiter, Johannes; Ebenhoech, Stefan; Schott, Dominik; Reindl, L.; Koehler, Sven et al. (2017b): Analytical engineering process to identify, assess and improve technical resilience capabilities. In Marko Čepin, Radim Briš (Eds.): ESREL 2017 (Portoroz, Slovenia, 18–22 June, 2017). The 2nd International Conference on Engineering Sciences and Technologies. High Tatras Mountains, Tatranské Matliare, Slovak Republic, 29 June – 1 July 2016. Boca Raton: CRC Press, p. 159.

Häring, Ivo; Gelhausen, Patrick (2018): Technical safety and reliability methods for resilience engineering: Taylor and Francis Group, pp. 1253–1260. Available online at https://doi.org/10.1201/9781351174664, https://www.taylorfrancis.com/books/e/9781351174664, checked on 10/17/2019.

InnoTherMS (2020a): Innovative Predictive High Efficient Thermal Management System – InnoTherMS. AllFraTech – DE-FR German-French Alliance for Innovative Mobility Solutions (AllFraTech). Available online at https://www.e-mobilbw.de/fileadmin/media/e-mobilbw/Publikationen/Flyer/AllFraTech_Infoflyer.pdf, checked on 3/6/2020.

InnoTherMS (2020b): Innovative Predictive High Efficient Thermal Management System – InnoTherMS. German-French research project. Elektromobilität Süd-West. Available online at https://www.emobil-sw.de/en/innotherms, checked on 5/15/2020.

MERLIN (2020): Multimodale effiziente und resiliente Lokalisierung für Intralogistik, Produktion und autonome Systeme. Demonstration Project, Funded by: Ministerium für Wirtschaft, Arbeit und Wohnungsbau Baden-Württemberg, Ministerium für Wissenschaft, Forschung und Kunst Baden-Württemberg, Fraunhofer-Gesellschaft für angewandte Forschung, e.V., Albert-Ludwigs-Universität Freiburg. Edited by Sustainablity Center Freiburg. Available online at https://www.leistungszentrum-nachhaltigkeit.de/demoprojekte/merlin/, checked on 9/27/2020.

OCTIKT (2020): Ein Organic-Computing basierter Ansatz zur Sicherstellung und Verbesserung der Resilienz in technischen und IKT-Systemen. German BMBF Project. Available online at https://projekt-octikt.fzi.de/, checked on 5/15/2020.

Resilienzmaße (2018): Resilienzmaße zur Optimierung technischer Systeme, Resilience measures for optimizing technical systems, Research Project, Sustainabilty Center Freiburg, 2016–2018. Available online at https://www.leistungszentrum-nachhaltigkeit.de/pilotphase-2015-2018/resilienzmasse/, checked on 9/27/2020.

Tomforde, Sven; Gelhausen, Patrick; Gruhl, Christian; Häring, Ivo; Sick, Bernhard (2019): Explicit Consideration of Resilience in Organic Computing Design Processes. In: ARCS Workshop 2019 and 32nd International Conference on Architecture of Computing Systems. Joint Conference. Copenhagen, Denmark. Berlin, Germany: VDE Verlag GmbH, pp. 51–56.

Chapter 2
Technical Safety and Reliability Methods for Resilience Engineering

2.1 Overview

Resilience of technical and socio-technical systems can be defined as their capability to behave in an acceptable way along the timeline pre-, during, and post-potentially dangerous or disruptive events, i.e., in all phases of the resilience cycle and overall. Hence, technical safety and reliability methods and processes for technical safety and reliability are strong candidate approaches to achieve the objective of engineering resilience for such systems.

The argument is also expected to hold true when restricting the set of methods to classical safety and reliability assessment methods, e.g., classical hazard analysis (HA) methods, inductive failure mode and effects analysis (FMEA), deductive fault tree analysis (FTA), reliability block diagrams (RBDs), event tree analysis (ETA), and reliability prediction. Such methods have the advantage that they are typically already used in industrial research and development. However, improving the resilience of systems is usually not their explicit aim.

The present chapter covers how to allocate such methods to different resilience assessment, response, development, and resilience management work phases or tasks when engineering resilience from a technical perspective. To this end, several assessment and analysis schemes, as well as risk control and resilience enhancement process schemes, are employed, as well as the resilience or disruption response cycle. Each such concept and the related process can be considered as a dimension to be considered in the generation of risk control and resilience.

In particular, the resilience dimensions of risk management, resilience objectives, resilience cycle time phases, technical resilience capabilities, and system layers are used explicitly to explore their range of applicability. Also, typical system graphical modeling, hardware and software development methods are assessed to document the usability of technical reliability and safety methods for resilience analytics and technically engineering resilience.

© The Author(s), under exclusive license to Springer Nature Singapore Pte Ltd. 2021
I. Häring, *Technical Safety, Reliability and Resilience*,
https://doi.org/10.1007/978-981-33-4272-9_2

As argued on abstract level, when defining resilience of technical systems as a technical capability of being sufficient resilient to potentially disruptive events, technical safety and reliability methods and processes are strong candidate approaches to achieve the objective of technical resilience. For instance, if the grid power or telecommunication grid supply fails, a resilient technical system maintains all or at least the most relevant and possibly safety–critical functions.

This holds true when defining disruptive events to cover statistical failures, known and unknown (unexampled, black swan) failures, and systematic failures and also when restricting disruptive events to major events occurring suddenly or creepingly.

The objective of overall resilience can be defined as the property of technical systems to overall sufficiently prevent disruptive events, being protected from disruptive events, responding to disruptive events, and recovering from disruptive events. This definition uses the resilience dimension, resilience (often also termed catastrophe) management cycle, and most of its steps to define resilience. In particular, successful preparation will lead to more efficient other steps.

The chapter shows how resilience dimensions can be used to structure the challenge of achieving resilient systems. The main objective is to show how technical safety and reliability methods can be used to support the generation of technical resilience within each resilience dimensional attribute.

For instance, it is identified that the classical methods hazard list and hazard analysis are well suited to support all phases of the resilience cycle (catastrophe management cycle), with a focus on prevention and protection. However, if hazard events are defined properly, i.e., including also such events as being not capable to respond due to lack of technical capabilities, hazard analyses may also cover the response and recovery phase. In a similar way, other approaches of assessing, developing, and improving resilience of systems are shortly introduced. For each such resilience dimension, it is also proposed how processes (see Chaps. 10 and 11) and methods (see Chaps. 5 to 9, Chaps. 13, and 14) as introduced and described in the textbook can be employed.

For each resilience dimension, a tabular scheme is presented for identifying the most relevant processes and methods using the resilience dimension attributes. In each case, the allocation of methods and processes for the generation of resilience is motivated and examples are given for illustration. By considering more resilience dimensions, different perspectives on suitability of suitable methods are achieved.

Section 2.2 gives arguments why it is expected to be beneficial to leverage system analysis methods for engineering resilience. Section 2.3 introduces the notion of resilience engineering dimensions, which are used single or in combination to develop, retrofit, or optimize resilient systems. It also describes how the following sections allocate technical safety and reliability methods for resilience engineering using the dimensional approach. Sections 2.4 to 2.8 present tabular allocations using, respectively, a single resilience dimension and all the processes and methods covered in the textbook. Section 2.9 summarizes and concludes. It also indicates how to use more than two resilience engineering dimensions to identify methods relevant for resilience engineering.

Chapter 1 mainly rests on (Häring et al. 2016a, b) but mainly on (Häring and Gelhausen 2018). Further sources include (Häring et al. 2016d; 2017b).

2.2 Why to Leverage Classical System Analysis Approaches for Resilience Engineering

As the number of applications of the concept of resilience to (socio) technical systems in mainly academic research and development rises, the question of how to successfully implement these approaches in the private sector, industry, and small and medium enterprises is getting more and more prominent. The present chapter addresses this challenge by surveying the suitability of classical, mainly analytical system analysis methods for assessing and improving resilience of (socio) technical systems.

The wording resilience analytics has been used recently quite general in the sense of resilience assessment in the societal–technical domain, see, e.g., (Linkov et al. 2016; López-Cuevas et al. 2017; Thorisson et al. 2017). However, the present use of the term analytical is to relate to classical system analysis methods, such as hazard analyses (HAs) including hazard lists (HLs), failure mode and effect analyses (FMEAs), fault tree analysis (FTA), event tree analysis (ETA), reliability block diagrams (RBD), and double failure matrix (DFM).

The application of established system modeling, analysis, and simulation methods for resilience analysis covers Bayesian networks (e.g., Yodo et al. 2017) and Markov models (e.g., Zhao et al. 2017). Fault propagation in a hazard and operability analysis (HAZOP) context for resilience assessment is described in (Cai et al. 2015). Also, first approaches have been reported to use FMEA for resilience assessment with the aim of applying the functional resonance analysis methodology (FRAM) to a smart building (Mock et al. 2016). However, for instance, fault tree analysis has not yet been used for resilience analysis.

First attempts of an evaluation of the suitability of system analysis, and simulation and development methods including some selected classical system analysis methods for resilience assessment have been conducted in (Häring et al. 2016a, c, e). In contrast to (Häring et al. 2017a), the present approach does not provide generic considerations of the suitability of methods for technically driven resilience assessment and development process steps.

The present approach focuses on determining which contributions to resilience assessment can be expected from mainly analytical system analysis methods and their extensions. To this end, it resorts to already often used resilience dimensions such as resilience or catastrophe response phases, system management domains, or resilience capabilities.

By resorting to five such resilience dimensions as detailed and motivated below, the expected relevancy of mainly analytical system analysis methods is assessed from different and complementary perspectives. Using the resilience dimensions, the suitability of method assessment is resolving the expected benefit in respective phases or resilience aspects rather than aiming at an overall applicability scoring.

This is conducted based on expert judgment (Meyer and Booker 2001) and consensus feedback of scientists related to the research field of technical safety and risk analysis. Also, groups of almost finished master students in security and safety engineering of an applied science university contributed mainly trained in tabular system analysis methods.

Key motivations for focusing on classical system analysis methods include the following:

- Analytical system analysis methods are established and accepted by practitioners in industry.
- Expectation that resilience analysis can in parts be delivered with extensions of classical methods.
- Expected efficiency of semi-quantitative methods compared to quantitative approaches.
- Identification of implicit resilience activities within current existing risk analysis and management practice.
- Clarification of resilience concepts by specifying methods supporting their fulfillment.
- Identification of critical resilience aspects that need to be analyzed with more effort, i.e., going beyond classical system analysis methods.

The chapter is structured as follows. In Sect. 2.3, the approach is described how to assess the suitability of mainly classical analytical system analysis methods for resilience analysis by employing resilience dimensions suitable for technical resilience understanding. Section 2.3 also details the methodology. It illustrates the need of going beyond classical risk assessment with the help of resilience event propagation through logic and assessment layers.

In Sects. 2.4 to 2.8, for each of the listed resilience concepts, possible contributions from the methods are discussed. For each resilience dimension, a matrix is filled with assessments of the suitability of the method for contributing to each of the resilience dimension attribute. Also, recommendations for the extension of the methods are given.

In Sect. 2.8, the overall suitability of each method is summarized and conclusions regarding adaptations and further developments are drawn.

2.3 Approach to Assess the Suitability of Methods

Before detailing the approach of suitability assessment of classical system analysis methods, some general considerations are provided on the necessary extension of methods for resilience assessment when compared to classical risk assessment.

Conditional probability expressions based and extending classical notions of risk have recently been used to quantify key objectives of resilient response (Aven 2017).

This shows that resilience analysis may benefit from the application of traditional and more modern risk concepts.

The idea used as starting point in (Aven 2017) is that resilience behavior can be defined to occur post-disruption events. Thus, resilience event, e.g., "system stabilizes post disruption," "system recovers," "system bounces back better," "recovery time shorter than critical time," or "sufficient system performance level reached within t" are always conditional previous events,

$$P(B|A), \tag{2.1}$$

where A is a "disruptive event" or equals a chain of events,

$$A = A_1, A_2, \ldots, A_n. \tag{2.2}$$

This approach relates with the often used definition of conditional vulnerability in risk expressions, see, e.g., (Daniel M. Gerstein et al. 2016),

$$R = P(E)\ P(C|E)\ C, \tag{2.3}$$

where E is a threat event and C the consequence.

However, the classical vulnerability approach of (2.3) focuses on the quantification of the conditional consequence probability, whereas (2.1) refers to resilience behavior post-disruption events.

As the vulnerability including risk definition of (2.3) is already an extension of the classical definition of risk

$$R = P(C)\ C, \tag{2.4}$$

and typical resilience expressions are further extending the definition of (2.1) and (2.3), it is expected that classical system reliability and safety approaches are challenged when used for assessing resilience. In particular, (very) simple tabular approaches resort to risk concepts as described by (2.4) when applied in a traditional way, i.e., they focus on avoidance of events and system robustness in case of events only.

Generalizing (2.1), resilience expressions of interest typically are of the form (Häring et al. 2016a)

$$P(B|A) = \sum_{i=1}^{N} P(B|D_i)P(D_i|A), \tag{2.5}$$

where D_i, $i = 1, 2, \ldots, n$, form a complete set of expansion events. Equation (2.5) uses the law of total probability and can be understood as an insertion of unity of all possible intermediate states

$$\sum_{i=1}^{n} P(\cdot|D_i)P(D_i|\cdot) \tag{2.6}$$

between any two known states.

Equation (2.5) can also express the idea of possibly unknown transition states or disruptions which are included in the set D_i. In this case, A is just a system initial state. In addition, (2.5) can be generalized to consider multiple resilience layers or response and recovery phases, see (Häring et al. 2016a).

Along the lines of interpretation given for (2.1), an interpretation of (2.5) reads, for instance,

$A =$ "Disruption event",

$\{D_i\}_{i=1,...,n}$, Set of possible response and recovery

events/Set of transition states,

$B =$ "Final state of interest".   (2.7)

When comparing risk and vulnerability expressions of the form (2.3) and (2.4) with resilience expressions of the form (2.1) and (2.5), it becomes obvious that it is not straightforward to expect that classical analytical system analysis methods can deliver assessment results regarding resilience. This motivates the question how such methods can contribute to resilience assessment.

For focusing the research question, the following system modeling, classical system analysis, and system development methods are considered regarding their suitability for resilience assessment:

- SysML, UML;
- HL, PHA, HA, O&SHA, HAZOP;
- FMEA, FMECA, FMEDA;
- RBD;
- ETA;
- DFM;
- FTA, time-dependent FTA (TDFTA);
- Reliability prediction with standards;
- Methods for HW and SW development; and
- Bit error-correction methods.

To assess the suitability of methods for resilience engineering, the following resilience dimensions are used:

- Five-step risk management process (AS/NZS ISO 31,000:2009), for review: (Purdy 2010; Luko 2013), for critical discussion mainly regarding the coverage of uncertainty (Aven 2011).

- Resilience time-phase cycle, based on (Thoma 2014), similar to catastrophe management cycle.
- Technical resilience capabilities, based on (Häring et al. 2016a).
- System layers, based on (Häring et al. 2016a; Häring and Gelhausen 2018).
- Resilience criteria (Bruneau et al. 2003; Häring et al. 2016a; Pant et al. 2014), for instance, 4Rs, 6R.
- Resilience analysis and management process (Häring et al. 2016a).

For the first five resilience dimensions, each combination of system analysis method and resilience dimension attribute is assessed using the three equivalent semi-quantitative scales

$$\{1, 2, 3, 4, 5\},$$
$$\{--, -, o, +, ++\},$$
$$\{\text{not suited (adaptation useless)},$$
$$\text{rather not suited (or only with major modifications)},$$
$$\text{potentially suited (after adaptation)},$$
$$\text{suited (with minor modifications)},$$
$$\text{very well suited (straightforward/no adaptions)}\}. \tag{2.8}$$

Typical examples read as follows: (i) The identification of potential disruption events of systems can be supported by using the classical system analysis methods hazard list (HL) and preliminary hazard analysis (PHA). Hazard lists are very well suited for identifying hitherto unknown events when used as checklists of potential disruptions and asking the question "what if?". Regarding the identification of possible disruptions for a system under consideration, the overall rating of HL could be "++" or "+" for PHA. This example shows that rather than assessing the generic suitability of a method, its use within a certain resilience assessment process or conceptual structuring is addressed.

(ii) Fault tree analysis (FTA) allows to consider combinations of events by using the AND gate. When only a known sequence of events is possible, the sequencing AND gate can be used, which enforces an order of occurrence of events. Such a sequence might be first "detection of threat," second "decision to start countermeasure," and third "activation of countermeasure." This order is then used for assessing the probability of success of a technical prevention measure. Sequential events can be analyzed with time-dependent Boolean differences to analyze sequential structure functions rather than classical combinatorial Boolean structure functions (Moret and Thomason 1984). Hence, FTA and even more TDFTA can be expected to cover after modifications also the response and recovery phase, resulting in a "+" assessment, respectively.

2.4   Method Suitability Assessment with Five-Step Risk Management Scheme

The five-step risk management scheme is only a very generic framework for identifying risks on resilience objectives. As discussed in the introduction, objectives in the case of resilience analysis are more second order (e.g., "fast recovery in case of disruption") when compared to classical risk analysis and management (e.g., "avoid disruption").

Table 2.1 assesses the suitability of analytical system analysis and some development methods for resilience analysis sorted along the five-step risk management scheme using the scale of (2.8).

Table 2.1 Suitability of analytical system analysis and HW/SW development methods for resilience analysis sorted along the five-step risk management scheme

Method \ five-step risk management process steps	(1) Establish context	(2) Identify risk/ hazards	(3) Analyze/ compute risks	(4) Evaluate risks	(5) Mitigate risks
SysML	+	++	++	o	++
UML	+	+	+	o	+
HL	+	++	+	o	−
PHA	+	+	++	+	+
SSHA, HA	o	+	++	++	++
O&SHA, HAZOP	o	+	++	++	++
FMEA, FMECA	−	o	++	+	++
FMEDA	−	−	++	+	++
RBD	o	+	++	+	++
ETA	+	+	++	+	++
DFM	−	o	++	+	++
FTA, TDFTA	−	o	++	+	++
Reliability prediction with standards	−	−	++	+	++
Methods for HW and SW development	−	−	−	−	++
Bit error-correction methods	−	−	−	−	++

Understanding the system sufficiently for resilience risk analysis is supported with graphical/semi-formal unified/systems modeling languages (UML/SysML) modeling, see the first two lines of Table 2.1.

The initial hazard analysis methods HL and PHL support the identification of possible disruptions. They are considered as a starting point. Refined analyses can be supported with SSHA, HA, O&SHA, and HAZOP, the differences of which are typically small and depend on the application; for small systems, they can be summarized in one analysis.

Approaches that need substantial system knowledge include RBD, the inductive approaches ETA, FME(D/C)A, and deductive approaches (TD) FTA, which are often summarized in a bow-tie analysis. The success of FMEA variations is expected to be more efficient when depending on system functions (or services) as inductive starting points rather than system components or subsystems.

In the case of (TD) FTA, the success of application will strongly depend on the definitions of the top events, which should cover main resilience objectives.

2.5 Method Usability Assessment Using Resilience Response Cycle Time Phases

The catastrophe management cycle in four steps (e.g., preparation, prevention and protection, response and recovery, learning and adaption) as well as in five steps as shown in Table 2.2 takes advantage of a logic or time ordering of events with respect to disruption events (Häring et al. 2016a): (far) before, during, immediately (after). Another typical timeline as well as logic sequence example was given in Sect. 2.5.

The first observation in Table 2.2 is that the analysis methods should be conducted, if considered relevant, mainly in the preparation phase. However, especially fast analytical simple methods can also be applied during actual conduction of response and recovery. For instance, during and post events, a PHA scheme could be used to identify further possible second-order events given a disruption.

The second observation in Table 2.2 comprises the coverage of resilience cycle phases. The suitability of method assessment stems from the fact that classical analytical approaches by definition cover prevention and protection when identified with frequency of event assessment and immediate (first order) damage assessment.

If system failure, in case of variations of HA, FMEA, and FTA, is defined as failure of adequate response (e.g., absorption and stabilization), of recovery (e.g., reconstruction and rebuilding) or even of improving or bouncing forward using damage as optimization opportunity, these methods can be used with adaptions also for these resilience timeline phases. Similarly, RBD and ETA are assessed.

The system modeling and development methods for hardware and software (HW/SW) can be used for all resilience cycle phases. As in the case of classical system analysis methods, adaptions up to major new developments are believed to be necessary.

Table 2.2 Suitability of system modeling, analytical system analysis, and selected development methods for resilience analysis along the five phases of the resilience cycle: resilience event order logic or timeline

Method \ Resilience timeline cycle phase	(1) Prepare	(2) Prevent	(3) Protect	(4) Respond	(5) Recover
SysML	++	++	++	++	++
UML	++	+	+	+	+
HL	++	++	++	++	++
PHA	++	++	++	+	+
SSHA, HA	++	++	++	+	+
O&SHA, HAZOP	++	++	++	+	+
FMEA, FMECA	++	++	++	+	+
FMEDA	++	++	++	+	+
RBD	++	++	++	+	+
ETA	++	++	++	+	+
DFM	++	++	++	+	+
FTA, TDFTA	++	++	++	+	+
Reliability prediction with standards	++	++	++	+	+
Methods for HW and SW development	++	++	++	+	+
Bit error-correction methods	++	++	++	++	+

Even if especially the classical tabular system analysis methods were assessed as very relevant for resilience assessment, it is noted that they are part of established processes in practice. Therefore, even when adding only some additional columns, their modified best practice of use is expected to be challenging in company development environments. In this sense, Table 2.2 is a guideline for the expected usability of the listed methods.

2.6 Method Usability Assessment Using Technical Resilience Capabilities

Sensor-logic-actor chains are basic functional elements used for active safety applications in safety instrumented systems, especially within the context of functional safety as governed by (IEC 61,508 Series). The technical resilience capabilities can be considered as a generalization of such functional capabilities.

They can also be related to the much more abstract and generic OODA (observe, orient, decide, act) loop, which has found much application also in the catastrophe response arena, see, e.g., (Lubitz et al. 2008; Huang 2015).

The technical resilience capabilities are also very close to capabilities to be expected from a general artificial intelligence (Baum et al. 2010) and related possible architectures (Goertzel et al. 2008).

Table 2.3 assesses the suitability for use of the selected methods along each technical resilience capability dimension attribute. Since the technical resilience capabilities are generic properties of (socio) technical systems, the realization of the properties in systems is prone to risks, e.g., external and internal; accidental and intentional; safety and security related; and systematic (by construction) and statistic.

In Table 2.3, the more generic system modeling methods SysML and UML are rated better when compared to more specific methods. Table 2.3 expresses with the uniform distribution of "+" that any resilience analysis conducted using the methods has to consider all the technical resilience properties. This shows that major adaptations and further developments are necessary to apply classical methods, since a cross-cutting task has to be covered by the methods.

For instance, columns or labels could be added to assess to which type of system resilience function a system failure belongs in case of HA, FMEA, and ETAs. Also, FTAs top-level event formulations either have to address the functional steps separately or find sufficient generic top-level formulations allowing for combinations of top events.

2.7 Method Usability Assessment Using System Layers

Table 2.4 assesses the potential of application of the representative methods of Sect. 2.3 with the help of system layers for socio-technical systems. The often used four layers, physical, information, cognitive, and social, see, e.g., (Fox-Lent et al. 2015), have been refined in the physical–technical domain and more specified in all attributes when compared to (Häring et al. 2016a).

The strength of the selected representative very specific methods is in the domain of hardware and data integrity as well as HW/SW development. Also, the classical tabular methods focus somewhat on electronics, especially FMEDA.

The general-purpose methods ETA, DFM, RBD, and FTA require educated application, often out of their classical domain of application, especially HAZOP. RBD diagrams and SysML/UML methods are expected to be of use for acquiring and documenting sufficient system understanding.

HA-type methods are well suited but need to be applied out of their typical technical domain also to operational and societal system layers.

2.8 Method Usability Assessment Using Resilience Criteria

Table 2.5 assesses the potential of application of the representative methods of Sect. 2.3 with the help of the often used four resilience criteria introduced by (Bruneau

Table 2.3 Suitability of system modeling, analytical system analysis, and selected development methods for resilience analysis along the technical resilience capability dimension attributes

Method\Technical resilience capabilities	(1) Observation, sensing	(2) Representation, modeling, Simulation	(3) Inference, decision-making	(4) Activation, action	(5) Learning, modification, adaption, rearrangement
SysML	++	++	++	++	++
UML	+	++	++	+	++
HL	+	+	+	+	+
PHA	+	+	+	+	+
HA	+	+	+	+	+
O&SHA, HAZOP	+	+	+	+	+
FMEA, FMECA	+	+	+	+	+
FMEDA	+	+	+	+	+
RBD	+	+	+	+	+
ETA	+	+	+	+	+
DFM	+	+	+	+	+
FTA, TDFTA	+	+	+	+	+
Reliability prediction with standards	+	+	+	+	+
Methods for HW and SW development	+	+	+	+	+
Bit error-correction methods	o	o	o	o	o

Table 2.4 Suitability of system modeling, analytical system analysis, and selected development methods for resilience analysis along system layers or generic management domains

Method\ System layer, management domain	(1) Physical	(2) Technical, hardware	(3) Cyber, software-wise,.protocols	(4) Operational, organizational	(5) Societal, economic, ethical
SysML	++	++	++	+	+
UML	++	++	++	+	+
HL	+	++	+	o	o
PHA	+	++	+	o	o
HA	+	++	+	o	o
O&SHA, HAZOP	+	++	+	+	+
FMEA, FMECA	+	++	+	o	o
FMEDA	+	++	+	o	o
RBD	+	++	+	o	o
ETA	+	++	+	+	+
DFM	+	++	+	+	+
FTA, TDFTA	+	++	+	o	o
Reliability prediction with standards	–	++	–	o	o
Methods for HW and SW development	+	++	++	o	o
Bit error-correction methods	–	+	++	o	o

et al. 2003) and technically refined by (Pant et al. 2014). For the suitability of method assessment, the following modified working definitions are used in this chapter:

(1) Robustness: measure for low level of damage (vulnerability) in case of event; "good" absorption behavior; "good" protection.

(2) Redundancy: measure for low level of overall system effect in case of local (in space, in time, etc.) disruption event; system disruption tolerance.

(3) Resourcefulness: measure for capability of successful allocation of resources in the response phase to stabilize the system post disruptions.

(4) Rapidity: measure for fast recovery of system.

The results are similar to the suitability assessment along the logic or timeline resilience cycle phases as conducted in Sect. 2.5: classical analytical approaches do not focus beyond the damage events. Robustness, resourcefulness, and rapidity are

Table 2.5 Suitability of system modeling, analytical system analysis, and selected development methods for resilience analysis along modified resilience criteria

Method \ Modified resilience criteria	(1) Robustness: low initial damage	(2) Redundancy: system property of overall damage tolerance	(3) Resourcefulness: fast stabilization and response	(4) Rapidity: fast recovery and reconstruction
SysML	+	++	++	++
UML	o	++	+	+
HL	+	o	−	−
PHA	+	+	−	−
HA	+	+	−	−
O&SHA, HAZOP	−	+	−	−
FMEA, FMECA	+	o	−	−
FMEDA	o	−	−	−
RBD	+	++	−	−
ETA	+	+	−	−
DFM	o	++	−	−
FTA, TDFTA	+	++	+	+
Reliability prediction with standards	−	−	−	−
Methods for HW and SW development	−	−	−	−
Bit error-correction methods	−	−	−	−

according to the working definitions strongly related to the resilience cycle phases, absorption/protection, response, and recovery.

Redundancy is understood in the classical way as an overall system property. Hence, in all cases, sufficient system understanding is required, which is supported by graphical modeling.

The (also) graphical approaches RBD, ETA, and FTA are strong for redundancy and resourcefulness assessment. Especially time-dependent FTA and underlying time-dependent Markov models are believed to be key for resourcefulness and redundancy assessment, nevertheless with major adaptations.

2.9 Conclusions Regarding Selection and Adoption of Methods

In summary, each of the representative analytical system analysis methods as well as HW/SW development methods (techniques and measures in the sense of (IEC 61,508 Series)) showed potential for resilience engineering, i.e., resilience assessment and development and optimization as defined in the introductory sections.

The classical tabular approaches HA and FMEA are assessed to be suited with minor up to major modifications for resilience analytics. Major advantages are expected by redefining and adding dedicated columns to cover resilience aspects. Also, graphical methods like RBD, ETA, and FTA are tools that by definition cover at least technical aspects of resilience of systems in case of very informed application.

In all cases, the extensions and adaptions need to carefully consider the initial background and application context of the methods. Therefore, in case of technical resilience engineering contexts, it is expected that the methods have to be newly established. This holds since all these methods are prone to routinely use, which is often very contrary to the out-of-the-box thinking necessary for resilience engineering. For instance, established hazard lists for an application domain will not contribute to an as complete as possible disruption threat list.

The different resilience dimensions used for suitability assessment exhibited strengths and weaknesses for exploring the methods' potentials:

- The risk management cycle is a very generic process, allocating most analysis methods in the risk analysis step. Resilience objectives formulation is key and challenge.
- Resilience cycle (time or logic) phases allow to spread out assessments and activities. However, they are prone to "divide et impera" effects of losing the overall picture.
- Technical resilience capabilities need to be covered for the operation of typical system (service) functions on overall system level allowing a technical approach. It is deemed challenging how to modify and extend classical methods to cover them.

Traditional resilience criteria ("Resilience Rs") can be nicely linked to timeline/logic concepts as well as system redundancy assessments. They also link with performance-based resilience curve assessments. Challenges are expected when trying to translate the more abstract concepts into system analysis and development requests.

In summary, future work is expected to benefit from informed further development of classical system analysis methods for resilience analysis. Such resilience analytics is believed also to strongly support the development of resilient systems, in particular, in industrial environments. Such informed applications are expected to ripe many of the benefits listed in the bullet list Sect. 2.2.

2.10 Questions

(1) Which resilience analysis phases, concepts, and dimensions do you know?
(2) How can the suitability of a classical technical safety analysis method be assessed regarding their use for resilience engineering?
(3) What are the main results regarding the potential use of classical system analysis methods for resilience engineering of socio-technical systems?

2.11 Answers

(1) See column headlines of Tables 2.1, 2.2, 2.3, 2.4, 2.5 and respective short explanations in text.
(2) There are several options, including (i) use of five-step risk management scheme of ISO 31,000 to identify for each step which classical methods can support the control of risks on resilience, e.g., use hazard list to identify potential disruptions of the system; (ii) use resilience cycle steps and ask which classical method is suited to support the step, e.g., use FTA to identify the combination of events that leads to failed evacuation of people in case of emergency, use SysML timing diagram to show the time constraints for successful evacuation, use event try to identify possible event trajectories post crisis event; and (iii) use system resilience properties to identify any weaknesses of socio-technical system under consideration with the help of SysML use case or activity diagram.
(3) See Sects. 2.2 and 2.9.

References

Aven, Terje (2011): On the new ISO guide on risk management terminology. In *Reliability Engineering and System Safety* 96 (7), pp. 719–726. https://doi.org/10.1016/j.ress.2010.12.020.
Aven, Terje (2017): How some types of risk assessments can support resilience analysis and management. In *Reliability Engineering & System Safety* 167, pp. 536–543. https://doi.org/10.1016/j.ress.2017.07.005.
Baum, Eric; Hutter, Marcus; Kitzelmann, Emanuel (Eds.) (2010): Artificial general intelligence. Proceedings of the Third Conference on Artificial General Intelligence, AGI 2010, Lugano, Switzerland, March 5–8, 2010. Conference on Artificial General Intelligence; AGI. Amsterdam: Atlantis Press (Advances in intelligent systems research, 10).
Bruneau, Michel; Chang, Stephanie E.; Eguchi, Ronald T.; Lee, George C.; O'Rourke, Thomas D.; Reinhorn, Andrei M. et al. (2003): A Framework to Quantitatively Assess and Enhance the Seismic Resilience of Communities. In *Earthquake Spectra* 19 (4), pp. 733–752. https://doi.org/10.1193/1.1623497.
Cai, Zhansheng; Hu, Jinqiu; Zhang, Laibin; Ma, Xi (2015): Hierarchical fault propagation and control modeling for the resilience analysis of process system. In *Chemical Engineering Research and Design* 103, pp. 50–60. https://doi.org/10.1016/j.cherd.2015.07.024.

Daniel M. Gerstein; James G. Kallimani; Lauren A. Mayer; Leila Meshkat; Jan Osburg; Paul Davis et al. (2016): Developing a Risk Assessment Methodology for the National Aeronautics and Space Administration. RAND Corporation (RR-1537-NASA). Available online at https://www.rand.org/pubs/research_reports/RR1537.html.

Fox-Lent, Cate; Bates, Matthew E.; Linkov, Igor (2015): A matrix approach to community resilience assessment. An illustrative case at Rockaway Peninsula. In *Environ Syst Decis* 35 (2), pp. 209–218. https://doi.org/10.1007/s10669-015-9555-4.

Goertzel, Ben; Wang, Pei; Franklin, Stan (Eds.) (2008): Artificial general intelligence, 2008. Proceedings of the First AGI Conference. ebrary, Inc; AGI Conference. Amsterdam, Washington, DC: IOS Press (Frontiers in artificial intelligence and applications, v. 171).

Häring, Ivo; Ebenhöch, Stefan; Stolz, Alexander (2016a): Quantifying resilience for resilience engineering of socio technical systems. In *Eur J Secur Res (European Journal for Security Research)* 1 (1), pp. 21–58. https://doi.org/10.1007/s41125-015-0001-x.

Häring, Ivo; Scharte, Benjamin; Hiermaier, Stefan (2016b): Towards a novel and applicable approach for Resilience Engineering. In Marc Stal, Daniel Sigrist, Susanne Wahlen, Jill Portmann, James Glover, Nikki Bernabe et al. (Eds.): 6-th International Disaster and Risk Conference: Integrative Risk Management – towards resilient cities // Integrative risk management - towards resilient cities. IDRC. Davos, 28.08-01.09.2016. Global Risk Forum, GRF. Davos: GRF, pp. 272–276. Available online at https://idrc.info/fileadmin/user_upload/idrc/proceedings2016/Extended%20Abstracts%20IDRC%202016_final2408.pdf.

Häring, Ivo; Scharte, Benjamin; Hiermaier, Stefan (2016c): Towards a novel and applicable approach for Resilience Engineering. In: 6-th International Disaster and Risk Conference (IDRC). Integrative Risk Management – towards resilient cities.

Häring, Ivo; Scharte, Benjamin; Stolz, Alexander; Leismann, Tobias; Hiermaier, Stefan (2016d): Resilience engineering and quantification for sustainable systems development and assessment: socio-technical systems and critical infrastructures. In Igor Linkov, Marie-Valentine Florin, Benjamin Trump (Eds.): IRGC Resource Guide on Resilience, vol. 1. Lausanne: International Risk Governance Center (1), pp. 81–89. Available online at https://irgc.org/risk-governance/resilience/irgc-resource-guide-on-resilience/volume-1/, checked on 1/4/2010.

Häring, Ivo; Scharte, Benjamin; Stolz, Alexander; Leismann, Tobias; Hiermaier, Stefan (2016e): Resilience Engineering and Quantification for Sustainable Systems Development and Assessment: Socio-technical Systems and Critical Infrastructures. In: Resource Guide on Resilience. Lausanne: EPFL International Risk Governance Center.

Häring, Ivo; Sansavini, Giovanni; Bellini, Emanuele; Martyn, Nick; Kovalenko, Tatyana; Kitsak, Maksim et al. (2017a): Towards a generic resilience management, quantification and development process: general definitions, requirements, methods, techniques and measures, and case studies. In Igor Linkov, José Manuel Palma-Oliveira (Eds.): Resilience and risk. Methods and application in environment, cyber and social domains. NATO Advanced Research Workshop on Resilience-Based Approaches to Critical Infrastructure Safeguarding. Dordrecht: Springer (NATO science for peace and security series. Series C, Environmental security), pp. 21–80. Available online at https://www.springer.com/de/book/9789402411225.

Häring, Ivo; Scheidereiter, Johannes; Ebenhoech, Stefan; Schott, Dominik; Reindl, L.; Koehler, Sven et al. (2017b): Analytical engineering process to identify, assess and improve technical resilience capabilities. In Marko Čepin, Radim Briš (Eds.): ESREL 2017 (Portoroz, Slovenia, 18–22 June, 2017). The 2nd International Conference on Engineering Sciences and Technologies. High Tatras Mountains, Tatranské Matliare, Slovak Republic, 29 June–1 July 2016. Boca Raton: CRC Press, p. 159.

Häring, Ivo; Gelhausen, Patrick (2018): Technical safety and reliability methods for resilience engineering. In S. Haugen, A. Barros, C. van Gulijk, T. Kongsvik, J. E. Vinnene (Eds.): Safety and Reliability - Safe Societies in a Changing World. Safety and Reliability – Safe Societies in a Changing World, Proceedings of the 28-th European Safety and Reliability Conference (ESREL), Trondheim, Norway, 17–21 June 2018: CRC Press, pp. 1253–1260.

Huang, Yanyan (2015): Modeling and simulation method of the emergency response systems based on OODA. In *Knowledge-Based Systems* 89, pp. 527–540. https://doi.org/10.1016/j.knosys.2015.08.020.

Linkov, Igor; Florin, Marie-Valentine; Trump, Benjamin (Eds.) (2016): IRGC Resource Guide on Resilience. L'Ecole polytechnique fédérale de Lausanne (EPFL) International Risk Governance Center. Lausanne: International Risk Governance Center (1). Available online at http://dx.doi.org/10.5075/epfl-irgc-228206, checked on 1/10/2019.

López-Cuevas, Armando; Ramírez-Márquez, José; Sanchez-Ante, Gildardo; Barker, Kash (2017): A Community Perspective on Resilience Analytics: A Visual Analysis of Community Mood. In *Risk analysis: an official publication of the Society for Risk Analysis* 37 (8), pp. 1566–1579. https://doi.org/10.1111/risa.12788.

Lubitz, D. K. von; Beakley, James E.; Patricelli, Frederic (2008): 'All hazards approach' to disaster management: the role of information and knowledge management, Boyd's OODA Loop, and network-centricity. In *Disasters* (32, 4), pp. 561–585. https://doi.org/10.1111/j.0361-3666.2008.01055.x.

Luko, Stephen N. (2013): Risk Management Principles and Guidelines. In *Quality Engineering* 25 (4), pp. 451–454. https://doi.org/10.1080/08982112.2013.814508.

Meyer, M. A.; Booker, J. M. (2001): Eliciting and Analyzing Expert Judgment. A Practical Guide: Society for Industrial and Applied Mathematics.

Mock, R.; Lopez de Obeso, Luis; Zipper, Christian (2016): Resilience assessment of internet of things. A case study on smart buildings. In Lesley Walls, Matthew Revie, Tim Bedford (Eds.): European Safety and Reliability Conference (ESREL). Glasgow, 25–29.09. London: Taylor & Francis Group, pp. 2260–2267.

Pant, Raghav; Barker, Kash; Ramirez-Marquez, Jose Emmanuel; Rocco, Claudio M. (2014): Stochastic measures of resilience and their application to container terminals. In *Computers & Industrial Engineering* 70, pp. 183–194. https://doi.org/10.1016/j.cie.2014.01.017.

Purdy, Grant (2010): ISO 31000:2009–Setting a new standard for risk management. In *Risk analysis: an official publication of the Society for Risk Analysis* 30 (6), pp. 881–886. https://doi.org/10.1111/j.1539-6924.2010.01442.x.

Thoma, Klaus (Ed.) (2014): Resilien-Tech: "Resilience by Design": a strategy for the technology issues of the future. München: Herbert Utz Verlag; Utz, Herbert (acatech STUDY).

Thorisson, Heimir; Lambert, James H.; Cardenas, John J.; Linkov, Igor (2017): Resilience Analytics with Application to Power Grid of a Developing Region. In *Risk analysis: an official publication of the Society for Risk Analysis* 37 (7), pp. 1268–1286. https://doi.org/10.1111/risa.12711.

Yodo, Nita; Wang, Pingfeng; Zhou, Zhi (2017): Predictive Resilience Analysis of Complex Systems Using Dynamic Bayesian Networks. In *IEEE Trans. Rel.* 66 (3), pp. 761–770. https://doi.org/10.1109/tr.2017.2722471.

Zhao, S.; Liu, X.; Zhuo, Y. (2017): Hybrid Hidden Markov Models for resilience metrics in a dynamic infrastructure system. In *Reliability Engineering & System Safety* 164, pp. 84–97. https://doi.org/10.1016/j.ress.2017.02.009.

Chapter 3
Basic Technical Safety Terms and Definitions

3.1 Overview

Well-defined key terms are crucial for developing safe systems. In particular, when developments are interdisciplinary with respect to (technical) disciplines involved, if more departments of a company participate, or different companies and industry sectors. This also holds true if originally different safety standard traditions are used, e.g., machine safety versus production process safety. Basic standards to be considered include ISO 31000 (2018), ISO 31010 (2019), and IEC 61508 (2010), where, in most cases, the terminology of the latter is taken up.

The chapter shows that a systematic ordering of key safety and reliability terms can be obtained along the following lines: Sorting of terms with respect to input terms needed by ascending from basic to higher level concepts. For example, from system, to risks of system, to acceptable risks, to risk reduction, to safety integrity level for functional safety, to safety-related system, etc.

Another ordering principle is to start out with basic terms before moving to definitions of time lines and processes, e.g., product life cycle, development life cycle, and safety life cycle.

The more general idea is that methods (techniques and measures including to some extent also organizational practices) can be understood as supporting and more often enabling well-defined processes. For instance, documenting the qualifications of employees is part of the prerequisites of successfully implementing a safety life cycle resulting in the development of safe and resilient systems. In a similar way, analytical tabular or graphical methods for determining safety requirements as well as the multiple methods for developing hardware and software sufficiently free from systematic and statistical errors can be understood as methods supporting processes.

In the context of technical safety, the chapter also distinguishes between system understanding, system modeling, system simulation, system analysis, and evaluation of system analysis results. It shows how to reserve the terms system simulation and system analysis to the application of the core methodologies, by excluding the preparation and modeling as well as the executive evaluations of the results of the

© The Author(s), under exclusive license to Springer Nature Singapore Pte Ltd. 2021   27
I. Häring, *Technical Safety, Reliability and Resilience*,
https://doi.org/10.1007/978-981-33-4272-9_3

methods employed. Of course, it emphasizes that a thorough analysis or simulation always comprises similar efforts for preparation and evaluation of results.

The chapter shows that this detailed separation of these, in practice, iteratively conducted working steps has the advantage to identify potentially identical steps when applying different simulation and analysis methods. In particular, deep system understanding as documented in appropriate system models is a candidate input for different analysis and simulation methods.

The chapter introduces first safety-related quantities (measures) and definitions, including quantitative risk, risk reduction, reliability on demand, reliability on high demand (continuous reliability), and safety integrity level (SIL) as a reliability quantity of safety functions.

For the cases of different level overlap of reliability-related and safety-related systems, the challenge of obtaining overall reliability and safety is addressed using a two-dimensional schematic. It can be used to select at top-level different safety design targets, in particular, separating or not separating safety-related functions from reliability functions.

Using the example step of system understanding and modeling, the chapter shows that the generation and assessment of technical safety and reliability strongly depends on the system type considered. This is reflected by the most appropriate modeling language for the system type developed, the granularity of the input information needed for the simulation or analysis method, and last but not least the information available for assessment. To this end, a number of system modeling languages are presented along with their typical application domain, focusing somewhat on electronic systems.

In this chapter, some basic terms of technical safety are introduced. They are defined and some brief explanation and examples are given.

The chapter starts with a list of basic terms. Sections 3.2–3.19 each explain one or two terms: system (simulation), safety (relevant systems and norms), safety function and integrity, (safety) life cycle, risk (minimization), hazard and reliability, as well as techniques and measures for achieving safety and models for the software and hardware development process. Additionally, system analysis methods are listed and forms of documentation given. The chapter ends with a set of questions in Sect. 3.20.

The chapter collects different definitions, mostly from IEC 61508 (2010) and as in parts reviewed in Siebold (2012). Further EMI sources are Heß et al. (2010), Mayrhofer (2010).

3.2   System

In the following, a *system* is understood as "a deterministic entity comprising an interacting collection of discrete elements" (Vesely et al. 1981).

Somewhat relaxing this classical definition, it is for the application of classical system analysis methods sufficient to replace "deterministic" by phrases like sufficiently understandable and accessible, e.g., by simulation or statistical approaches.

Sufficient in this context means that the information necessary as input for classical system analysis methods are available, e.g., possible system states and transitions.

The interactions of elements have to be seen as part of the system. They often are extremely complicated and an essential part of the system.

3.3   Life Cycle

The *life cycle* of a system is the time when the system is capable of being used (Branco Filho 2000). It contains all phases of the system from the conception to the disposal. These phases include (Blanchard and Fabrycky 2005) the following:

- the system conception,
- design and development,
- production and/or construction,
- distribution,
- installation/setup,
- operation (in all credible ways),
- maintenance and support,
- retirement,
- phase-out, and
- disposal.

The term life cycle also often only refers to the phases after the production.

3.4   Risk

Risk considers a measure for the frequency/probability of events and a measure for the consequences. Most definitions do not ask for a special relation between probability and consequences on the one hand and risk on the other hand. The classical definition of risk has the strong requirement of proportionality (Mayrhofer 2010):

"Risk should be proportional to the probability of occurrence as well as to the extent of damage." according to Blaise Pascal (1623-1662). Similar definitions have been used in many variations, see, for instance, the discussion in Kaplan and Garrick (1981), Haimes (2009).

Formalized this reads as follows.

Classical definition of risk: Risk is proportional to a measure for the probability P of an event (frequency, likelihood) and the consequences C of an event (impact, effect on objectives):

$$R = PC. \tag{3.1}$$

We work with this definition and generalizations thereof.

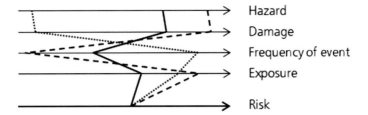

Fig. 3.1 Three examples of combinations of given parameters that can lead to a similar risk although they are very different. The dotted line can be associated with a car crash. Car crashes occur often but most of them do not have very big consequences. The dashed line on the contrary represents an event that occurs very seldom but has enormous consequences, for example, the explosion of a nuclear power plant. The straight line could, for example, stand for the collapse of a building with a medium probability of occurrence and a medium level of consequences

Figure 3.1 shows the meaning of the product in Eq. (3.1). It can be seen how different combinations of levels of hazard, damage, frequency, and exposure can all lead to the same risk.

3.5   Acceptable Risk

Acceptable risk is the risk that is tolerated by the society. Of course, the level of risk that is considered acceptable can be different for different events. The decision whether a risk is acceptable is not only based on physical or statistical observations but also on societal aspects.

Typically, the acceptance does not depend on R only but also on (P,C). This is further discussed in Sects. 7.9.4 and 8.10.2.

3.6   Hazard

By *hazard* we mean the physical hazard potential that propagates from a hazard source. Damage can occur if the hazard physically or temporally meets with objects or persons. Hence, hazard is not the same as unacceptable risk, as hazard does not include the frequency of hurting something or someone.

Remark Other sources often use hazard for a non-acceptable level of risk and safety for an acceptable level of risk. This differs from the definition of hazard used here.

3.7 Safety

The norm IEC 61508 defines *safety* as "freedom from unacceptable risks" (IEC 61508 2010). In the standard MIL-STD-882D, safety of a system is defined as "the application of engineering and management principles, criteria, and techniques to achieve acceptable mishap risk, within the constraints of operational effectiveness and suitability, time, and cost, throughout all phases of the system life cycle" (Department of Defense 2000; Mayrr et al. 2011).

Remark 1 Both definitions allow the risk to be greater than zero in a safe system. It only has to be sufficiently small.

Remark 2 All single risks must be acceptable as well as the total risk.

3.8 Risk Minimization

Risk minimization/risk reduction can be done by reducing the frequency of occurrence or the damage or both. For example, risk minimization can be achieved by inherent passive constructions, reactive and active safety functions, protective gear and safety-related information, or organizational measures (Neudörfer 2011).

Figure 3.2 illustrates risk minimization and relates the terms residual risk, tolerable risk, EUC (equipment under control) risk, and risk reduction to each other.

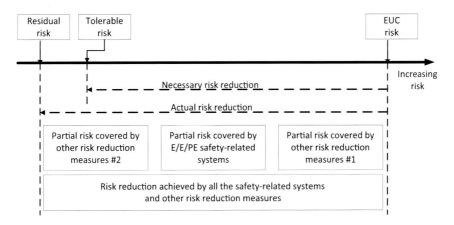

Fig. 3.2 Presentation of the correlation of the terms danger, risk, safety, and risk reduction from (IEC 61508 2010) with minor graphical modifications. Copyright © 2010 IEC Geneva, Switzerland. www.iec.ch

3.9   Safety-Relevant and Safety-Critical Systems

A system is *safety relevant* or *safety critical* if its risks to humans, the environment, and objects are only acceptable if at least a single technical subsystem is necessary to achieve an acceptable (sufficiently low) overall risk level. This can be achieved, in particular, by safety functions.

Remark 1 With this definition a safety-relevant or safety-critical system stays safety relevant if its risk has been reduced by risk-minimizing measures. The classification depends on the level of frequency and the level of damage.

Remark 2 Possible distinctions between safety relevant and safety critical include the level of risk that is controlled by safety functions as well as whether persons might be severely injured or not, or even if fatalities are feasible.

3.10   Safety-Relevant Norms

The application of standards and norms in the development of safety-relevant systems can be essential to achieve a tolerable level of risk for the system.

Examples

– An important norm in this context is the basic norm on functional safety IEC 61508 (DIN IEC 61508 2005; IEC 61508-SER 2005; IEC 61508 S+ 2010).
– Several norms are derived from the IEC 61508, for example, the ISO 26262 (2008) for the automotive sector (Siebold 2012).
– The railway system has its own norms and standards, for example, EN 50126 (1999) for safety and reliability (Siebold 2012).

3.11   Systems with High Requirements for Reliability

Safety is the absence of intolerable risks, see Sect. 3.7. This implies that in a safe system the components only are active in ways they are supposed to be and at times when they should be active. *Reliability*, on the other hand, means that components are always active when they are supposed to be and at every time they are supposed to be active.

Example An airbag in a car that cannot be started at all is safe but not reliable. Airbags that start when being initiated but also due to strong mechanical vibrations due to rough roads are reliable but not safe. Only airbags that start when they are initiated and only when they are lit are safe and reliable.

A common method to improve the safety and/or reliability of a system if the requirements are not fulfilled is to install redundancies, that is, additional parts in a system that are not needed if the system works without any failures.

3.12 Models for the Software and Hardware Development Process

Several well-known models such as the waterfall model, the spiral model, and the "V" model are used to model the software and/or hardware development process of a system. Those three listed examples will be introduced in Chap. 10 and their properties will be explained.

3.13 Safety Function and Integrity

A *safety function* is defined as "a function to be implemented by an *E/E/PE* [electrical/ electronic/ programmable electronic] safety-related *system* or other risk reduction measures, that is intended to achieve or maintain a safe state for the safety function, in respect of a specific hazardous event" (IEC 61508 2010).

According to IEC 61508, the reliable functioning of safety functions ensures the safety of a system. The safety functions are implemented by safety-related systems. In short, the safety-related systems, which are possibly identical to the overall system, have to be reliable to ensure safety of the system.

Figure 3.3 shows the different ways how the safety-related system can be combined with the original system.

Tables 3.1, 3.2, 3.3 and 3.4 show the reliability of the total system for the different possibilities to combine a system with a safety-related system from Fig. 3.3.

A *safety integrity level* (SIL) describes the probability that a safety function fails within a given period of time or amount of demands. There are two types of SILs

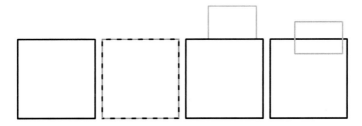

Fig. 3.3 From left to right: A system without safety-related system, a safety-related system that is included in the system, a separate safety-related system, and a safety-related system that is partly included in the system but partly outside the system

Table 3.1 The reliability of the total system for a system without safety-related system. The entries in the matrix show the reliability of the total system (reliability of system functions taking account of the safety function effects)

Reliability of safety system	Reliability of the system	
	Low	High
Low	Low	High
High	Low	High

Table 3.2 The reliability of the total system for a system with integrated safety-related system

Reliability of safety system	Reliability of the system	
	Low	High
Low	Low	Low
High	Low	High

Table 3.3 The reliability of the total system for a system with separate safety-related system

Reliability of safety system	Reliability of the system	
	Low	High
Low	Low	High
High	Depends	High

Table 3.4 The reliability of the total system for a system with partly integrated safety-related system

Reliability of safety system	Reliability of the system	
	Low	High
Low	Low	Depends
High	Depends	High

depending on the mode of operation of the safety function. On the one hand, a SIL for requirements with low demand rate and, on the other hand, a SIL for modes of operation with high demand rate or continuous mode, see Table 3.5.

Table 3.5 The SILs for high and low demand rates as defined in IEC 61508 (2010). Copyright © 2010 IEC Geneva, Switzerland. www.iec.ch

Low demand rate		High demand rate	
Per request		Per hour	
SIL	Average probability of a dangerous failure	SIL	Average frequency of a dangerous failure
4	$[10^{-5}, 10^{-4})$	4	$[10^{-9}h^{-1}, 10^{-8}h^{-1})$
3	$[10^{-4}, 10^{-3})$	3	$[10^{-8}h^{-1}, 10^{-7}h^{-1})$
2	$[10^{-3}, 10^{-2})$	2	$[10^{-7}h^{-1}, 10^{-6}h^{-1})$
1	$[10^{-2}, 10^{-1})$	1	$[10^{-6}h^{-1}, 10^{-5}h^{-1})$

The SILs of the safety functions of a system are determined by quantitative or qualitative analyses. For more detailed information, see IEC 61508 (2010).

3.14 Safety Life Cycle

The *Safety Life Cycle* described in IEC 61508 represents the life cycle of systems under development. It can be seen as a systematic concept to gain safety within the development process of a system. The application of the norm is a way to ensure the client that the system is safe. The safety life cycle will be explained in more detail in Chap. 11.

3.15 Techniques and Measures for Achieving Safety

Techniques and measures for achieving safety comprise organizational measures, system analysis methods, testing, software analysis, and others. They are mainly applied within the development and production phase. Some systems require their application also throughout the remaining life cycle, e.g., maintenance and tests.

3.16 System Description, System Modeling

Verbose system descriptions are one possible way of documenting system understanding. Often ad hoc pictures and schemes are added. Such descriptions are often very intuitive and allow for a description intelligible to all. However, only a certain level of rigor allows for (less) ambiguous system descriptions. This holds true even more when system models are used as input for (semi-)automatic system simulation and analyses.

Especially for the development of safety-critical systems, a consistent syntax in the documentation is very useful (Siebold 2012). To this end, this section lists different modeling languages.

Most of the languages listed below can be used at various levels of abstraction. Hence, also for overall system descriptions in early stages of the system development, even with rather little information on the system designs, corresponding models are feasible.

In general, it is advisable to select modeling languages that can be used for many or even during all development phases. By definition, SysML is such a systems modeling language. However, in practice, each technical discipline and often also technology has its own most established modeling language. Examples are UML for software developments and VHDL for electronic hardware developments.

3.16.1   OPM (Object Process Methodology)

The *object process methodology* (OPM) is a holistic approach to model complex systems where objects, processes, and states are the main modeling elements (Dori 2002; Siebold 2012).

In 2007, Gobshtein et al. declare OPM as one of the »state-of-the-art« system modeling languages (Grobshtein et al. 2007; Siebold 2012). The OPM uses two forms of representation of systems: the object process diagrams (OPD) to display systems graphically and the object process language (OPL) to display them textually (Grobshtein et al. 2007; Siebold 2012).

3.16.2   AADL (Architecture Analysis and Design Language)

The Society of Automotive Engineers (SAE) released the standard AS5506 called *Architecture Analysis and Design Language* (SAE Aerospace 2004). This system modeling language supports the system development in early estimates regarding performance, schedulability, and reliability (Feiler et al. 2006). AADL is good to model system which have strong requirements concerning space, storage, and energy but not so good to document a detailed design or the implementation of components (SAE Aerospace 2004; Siebold 2012).

3.16.3   UML (Unified Modeling Language)

The first draft of the *unified modeling language* (UML) was first presented to the object management group (OMG) in 1997 and can be seen as the lingua franca of software development (Rupp et al. 2007; Siebold 2012).

For modeling with UML, several different diagrams can be used. They will be explained in more detail in Chap. 13.

3.16.4   AltaRica

The modeling language *AltaRica* was developed by the Laboratoire Bordelais de Recherche en Informatique (LaBRI) and mostly consists of hierarchically ordered knots which can include states and state transitions (Griffault and Point 2006; Siebold 2012).

AltaRica Data Flow (AltaRica DF) is a dialect of AltaRica which is widely spread among reliability engineers (Cressent et al. 2011; Siebold 2012).

3.16.5 VHDL (Very High-Speed Integrated Circuit Hardware Description Language)

VHDL was created in the 1980 s as a documentation and simulation language and was first standardized in 1987 as IEEE 1067-1987 (Reichardt 2009; Siebold 2012). The language can be used in the design, description, simulation, testing, and a hierarchical specification of hardware (Navabi 1997).

3.16.6 BOM (Base Object Model)

The specification of the "Base Object Model" was accepted as a standard by the Simulation Interoperability Standards Organization (SISO) in 2006 (S.B.O.M.P.D. Group 2006; Simulation Interoperability Standards Organization 2011; SimVentions 2011; Siebold 2012). BOM contains different components to model different aspects of a system. The specification contains meta-information about the model, "Pattern of Interplay" tables to display recurring action patterns, state diagrams, entities, and events (Siebold 2012).

3.16.7 SysML (Systems Modeling Language)

The *systems modeling language* (SysML) is a semi-formal systems modeling language based on UML. It consists of some of the UML diagrams, some modified UML diagrams and two new diagram types. It will be explained in more detail in Chap. 14.

3.17 System Simulation

In the context of technical safety, system simulation is based on system models. System simulation comprises the static and dynamic behavior of systems. Corresponding to the system modeling which ranges from abstract (semi) formal models, to engineering models, e.g., circuit simulation, to numerical (coupled) simulation, very different types of simulation are feasible.

It is very common to use *numerical simulation* in the development process of a system. In general, the term "numerical simulation" describes methods to approximate the solution of complex physical problems for which often manageable analytic solutions do not exist. In this way, one can, for example, analyze the effects of design or environment on the system without having to test a real system. This is more time- and cost-efficient (Heß et al. 2010).

Example Heß et al. (2010): One simulation tool is "COMSOL Multiphysics" which is based on the *"Finite Element Method"* (FEM) (Comsol 2013). Its strength lies in solving coupled physical systems. It also has a graphical user interface where geometry and boundary conditions can be easily defined.

3.18 System Analysis Methods

System analysis methods are used to systematically gain information about a system (Vesely et al. 1981).

There are two different concepts for system analysis: inductive and deductive methods. *Inductive* system analysis methods analyze in which way components can malfunction and the consequences on the whole system. *Deductive* methods analyze how a potential failure of the whole system can occur and what has to happen for this to occur. Figure 3.4 visualizes the differences of inductive and deductive methods. The square represents the system.

In Sects. 3.16–3.18, the terms system modeling, system simulation, and system analysis were introduced. Their relation is visualized in Fig. 3.5. The modeled system can be used both for the system analysis and the system simulation. The results of both methods can go into one overall result.

Figure 3.6 visualizes the three terms system modeling, system simulation, and system analysis. The square represents the system. The first line shows the modeling of the system. The different parts of the system are described with a modeling language. The result is an overall description of the system. The modeled system

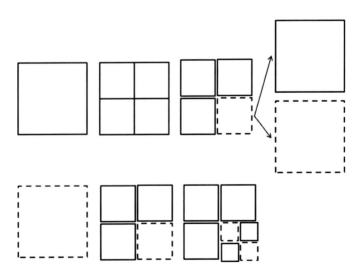

Fig. 3.4 Comparison of an inductive (top) and a deductive (bottom) approach

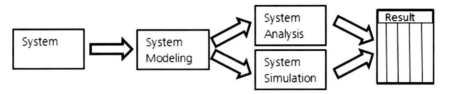

Fig. 3.5 Relation between the terms system modeling, system simulation, and system analysis

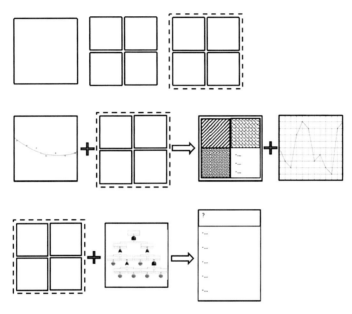

Fig. 3.6 Comparison of system modeling, system simulation, and system analysis (from top to bottom)

and some additional input or measurements can be used to simulate the system. Apart from the simulation of the system, there might be some further output. The modeled system can also be used as input for system analysis methods. One gets results about the safety and reliability of a system.

3.19 Forms of Documentation

Two forms of *documentation* of system analysis are widely used: tables and graphics.

Example FMEA is a method where results are documented in a table while the results of an FTA are documented in a graphic.

A problem is to find a good balance between the formalities of these forms of documentation and the content. Focusing too much on formalities ("to mistake form for substance" (Vesely et al. 1981)) yields less flexibility and not very deep results. Only focusing on the content, however, makes it hard for other persons to understand and process the results.

3.20 Questions

(1)   Which definitions do you need at least to define the safety of a system according to the functional safety approach? How are they related?
(2)   What is the difference between functional safety and more classical (non-functional) safety definitions?
(3)   Name three phases in the life cycle of a system.
(4)   What is the classical formula to compute risks?
(5)   What is the difference between hazard (hazard source, hazard potential), unacceptable, or high risk?
(6)   Relate the terms "residual risk," "tolerable risk," "necessary risk minimization," and "actual risk minimization" to each other.
(7)   Explain the difference between safety and reliability.
(8)   What is a safety life cycle?
(9)   What is a safety function according to IEC 61508?
(10) What does SIL stand for? How is it defined according to IEC 61508?
(11) Explain the difference between inductive and deductive system analysis methods.
(12) Give the upper and lower bounds for SIL 1 to SIL 4 functions in case of low and high demands. Illustrate them and motivate how they can be related to each other.

3.21 Answers

(1)   Technical terms used are: system, risk, acceptable risk, risk reduction, qualitative, and quantitative description of safety functions.
      In addition, when considering the technical implementation: EEPE-safety-related system, SIL, HFT, DC, and complexity level (A or B).
      A system is safe according to the functional safety approach, if all risks are known and reduced single and in total using appropriate risk reduction measures. At least one risk reduction measure of the system is qualitatively and quantitatively a safety function with at least SIL level 1 and realized as EEPE-safety-related system.
(2)   The functional safety approach requires that an EEPE-safety-related system function maintains actively a safe state of a system or mitigates a risk event to

an acceptable level during and post events. It asks to develop such functions, which can be described qualitatively and quantitatively.

(3) See the list in Sect. 3.3.

(4) $R = PC$, see Sect. 3.4.

(5) Hazard was defined as the hazard potential which may cause damage only if the hazard event occurs and objects are exposed, see Sect. 3.6. A non-acceptable risk comprises a non-acceptable combination of damage in case a hazard event occurs and its probability.

(6) Necessary risk minimization is the difference between the actual risk and the tolerable level of risk. The chosen actual minimization measures lead to a decrease of the risk to the residual risk. This is smaller than or equal to the tolerable risk. Hazard is often also used in the sense high-risk event. See Fig. 3.2.

(7) See Sect. 3.13.

(8) See Sect. 3.14.

(9) See Sect. 3.13.

(10) See Sect. 3.13. The safety integrity level (SIL) measures the reliability of the safety function.

(11) See Sect. 3.18. Inductive is also termed bottom-up and deductive top-down.

(12) See Table 3.5. When assuming that 10% probability of failure on demand translate to approximately a maximum failure frequency of $10\%/10000\,\text{h} = 1.0 \cdot 10^{-5}\text{h}^{-1}$, i.e., 10% failures per year, one has a relation for SIL 1. Similar assumptions are made for SIL 2 to SIL 4.

References

Blanchard, B. S. and W. J. Fabrycky (2005). Systems Engineering and Analysis, Prentice Hall.

Branco Filho, G., Confiabilidade e Manutebilidade (2000). Dicionário de Termos de Manutenção. Rio de Janeiro, Editora Ciência Moderna Ltda.

Comsol. (2013). "Multiphysics Modeling and Simulation Software." Retrieved 2013-03-08, from http://www.comsol.com/.

Cressent, R., V. Idasiak, F. Kratz and D. Pierre (2011). Mastering safety and reliability in a model based process. Reliability and Maintainability Symposium (RAMS).

Department of Defense (2000). MIL-STD-882D - Standard Practice For System Safety.

DIN IEC 61508 (2005). Funktionale Sicherheit sicherheitsbezogener elektrischer/elektronischer/programmierbarer elektronischer Systeme, Teile 0 - 7. DIN. Berlin, Beuth Verlag.

Dori, D. (2002). Object-Process Methodology: A Holistic Systems Pardigm. Berlin, Heidelberg, New York, Springer.

EN 50126 (1999). Railway applications. The specification and demonstration of reliability, availability, maintainability and safety (RAMS), Comité Européen deNormalisation Electrotechnique. **EN50126:1999**.

Feiler, P. H., D. P. Gluch and J. J. Hudak (2006). The Architecture Analysis & Design Language (AADL): An Introduction, Carnegie Mellon University.

Griffault, A. and G. Point (2006). "On the partial translation of Lustre programs into the AltaRica language and vice versa." Laboratoire Bordelais de Recherche en Informatique.

Grobshtein, Y., V. Perelman, E. Safra and D. Dori (2007). System Modeling Languages: OPM Versus SysML. International Conference on Systems Engineering and Modeling - ICSEM'07. Haifa, Israel, IEEE.

Haimes, Yacov Y. (2009): On the complex definition of risk: a systems-based approach. In *Risk Analysis* 29 (12), pp. 1647–1654. https://doi.org/10.1111/j.1539-6924.2009.01310.x.

Heß, S., R. Külls and S. Nau (2010). Theoretische und experimentelle Analyse eines neuartigen hochschockfesten Beschleunigungssensor-Designs. Diplom, Hochschule Ulm.

IEC 61508 (2010). Functional Safety of Electrical/Electronic/Programmable Electronic Safety-related Systems Edition 2.0 Geneva, International Electrotechnical Commission.

IEC 61508-SER (2005). Functional Safety of Electrical/Electronic/Programmable Electronic Safety-related Systems Ed. 1.0. Berlin, International Electrotechnical Commission, VDE Verlag.

IEC 61508 S+ (2010). Functional Safety of Electrical/Electronic/Programmable Electronic Safetyrelated Systems Ed. 2 Geneva, International Electrotechnical Commission.

ISO 26262 (2008). ISO Working Draft 26262: Road vehicles - Functional safety, International Organization for Standardization.

ISO 31000, 2018-02: Risk management - Guidelines. Available online at https://www.iso.org/standard/65694.html.

ISO 31010, 2019-06: Risk management - Risk assessment techniques. Available online at https://www.iso.org/standard/72140.html.

Kaplan, Stanley; Garrick, B. John (1981): On The Quantitative Definition of Risk. In *Risk Analysis* 1 (1), pp. 11–27. https://doi.org/10.1111/j.1539-6924.1981.tb01350.x.

Mayrhofer, C. (2010). Städtebauliche Gefährdungsanalyse - Abschlussbericht Fraunhofer Institut für Kurzzeitdynamik, Ernst-Mach-Institut, EMI, Bundesamt für Bevölkerungsschutz und Katastrophenhilfe, http://www.emi.fraunhofer.de/fileadmin/media/emi/geschaeftsfelder/Sicherheit/Downloads/FiB_7_Webdatei_101011.pdf.

Mayrr, A., R. Ploumlsch and M. Saft (2011). "Towards an operational safety standard for software: Modelling IEC 61508 Part 3." Proceedings of the 2011 18th IEEE International Conference and Workshops on Engineering of Computer Based Systems (ECBS 2011): 97-104104.

Navabi, Z. (1997). VHDL: Analysis and Modeling of Digital Systems, McGraw-Hill, Inc.

Neudörfer, A. (2011). Konstruieren sicherheitsgerechter Produkte, Springer.

Reichardt, J. (2009). Lehrbuch Digitaltechnik, Oldenbourg Wissenschaftsverlag GmbH.

Rupp, C., S. Queins and B. Zengler (2007). UML 2 Glasklar - Praxiswissen für die UML-Modellierung. München, Wien, Carl Hanser Verlag.

SAE Aerospace (2004). Architecture Analysis & Design Language (AADL), SAE Aerospace.

S.B.O.M.P.D. Group (2006). Base Object Model (BOM) Template Specification. Orlando, FL, USA, Simulation Interoperability Standards Organization (SISO).

Siebold, U. (2012). Identifikation und Analyse von sicherheitsbezogenen Komponenten in semiformalen Modellen. Doktor, Albert-Ludwigs-Universität Freiburg im Breisgau.

Simulation Interoperability Standards Organization. (2011). "Approved Standards." Retrieved 2011-10-31, from http://www.sisostds.org/ProductsPublications/Standards/SISOStandards.aspx.

SimVentions. (2011). "Resources - Technical Papers." Retrieved 2011-10-31, from http://www.simventions.net/resources/technical-papers.

Vesely, W. E., F. F. Goldberg, N. H. Roberts and D. F. Haasl (1981). Fault Tree Handbook. Washington, D.C., Systems and Reliability Research, Office of Nuclear Regulatroy Research, U.S. Nuclear Regulatory Comission.

Chapter 4
Introduction to System Modeling for System Analysis

4.1 Overview

System analysis methods are used to systematically gain information about a system (Vesely et al. 1981).

System analysis should start as early as possible because costs caused by unde-tected deficiencies increase exponentially in time (Müller and Tietjen 2003). An accompanying analysis during the development process of a product also helps the analyst to better understand the product which simplifies the detection of problems.

In this chapter, the general concepts and terms of system analysis are introduced which can be and should be applied in every system analysis method that is intro-duced in the following chapters. In Sects. 4.2 and 4.3, system and its boundaries are introduced. Section 4.4 explains why a theoretical system analysis is often more practicable than testing samples.

Sections 4.5 to 4.8 introduce terms to classify system analysis methods, e.g., to be able to address fault tree analysis as graphical deductive qualitative and quantitative methods. Sections 4.9 to 4.11 explain different types of failures in system analysis contexts. This is sufficient to introduce the terms reliability and safety in Sect. 4.12. System-redundancy-related terms are covered in Sects. 4.13 to 4.15. Section 4.16 illustrates the need of system analysis approaches for efficient system optimization and 4.17 stresses the need to cover failure combinations.

The chapter is a summary, translation, and partial paraphrasing of (Echterling et al. 2009) with some additional new examples and quotations.

4.2 Definition of a System

In the following, a *system* is understood as "a deterministic entity comprising an interacting collection of discrete elements" (Vesely et al. 1981).

© The Author(s), under exclusive license to Springer Nature Singapore Pte Ltd. 2021
I. Häring, *Technical Safety, Reliability and Resilience*,
https://doi.org/10.1007/978-981-33-4272-9_4

Determinacy is necessary in the definition because system analysis methods do not apply to an entity that makes its own decisions and the behavior of which cannot be determined from the outside. The interactions of elements have to be seen as part of the system. They are often extremely complicated and an essential part of the system.

However, this does not exclude the analysis of systems which are self-learning or autonomous, if such systems can be understood sufficiently in terms of their states and interactions, e.g., by model-based simulation or by boxing their behavior. Examples for such systems are, for instance, self-calibrating localization systems (ULTRA-FLUK 2020) and autonomous inspection systems of chemical plants (ISA4.0 2020).

4.3 Boundaries of the System

Before analyzing a system, the *boundaries of the system* have to be defined. One distinguishes between four different types of boundaries, namely,

– External boundaries,
– Internal boundaries,
– Temporal boundaries, and
– Probabilistic boundaries.

Although a clear allocation of one type is not always possible (Vesely et al. 1981; Bedford and Cooke 2001; Maluf and Gawadiak 2006).

External boundaries separate the system from the outside surroundings which cannot be or are not influenced by the system. For example, it can be useful to see the temperature of the surrounding area as part of the system and not part of the surroundings if it is influenced by the system.

Internal boundaries determine the resolution of a system. That is, they declare the different elements of a system. The resolution can be defined with respect to geometrical size, functionality, reliability, energy flow, and heat flow. The internal boundaries help to find a balance between the costs and the level of detail. The most detailed resolution of the system that can be regarded is not always the most efficient.

Temporal boundaries declare the length of time a system is used. Systems that are used for decades ask for a different system analysis than systems used for short time intervals.

Probabilistic boundaries are boundaries that are set because one already has sufficient information about the safety and reliability of certain parts of the system. In this case, it might be useful to regard such a part as one element in the analysis or to exclude it completely.

Figure 4.1 visualizes the different types of boundaries.

Deterministic behavior of systems may be achieved by the extension or refinement of system boundaries. For example, application software tools often receive updates that slightly modify their behavior or algorithms are often deterministic at a fine level of resolution.

Fig. 4.1 Visualization of the different boundary types

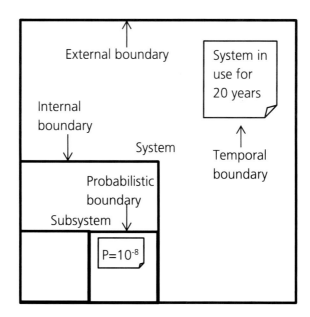

4.4 Theoretical Versus Practical System Analysis

We give an argument why theoretical system analysis often has to strongly complement or replace practical system tests. Suppose a system produces N parts with a
tolerance of K damaged parts. For assessing its reliability, the most obvious approach
is to test all N parts. However, in many cases this is not possible. If N is big, testing
all parts is very expensive and time-consuming. Another problem is that there are
products which are destroyed when being tested. In both cases, only a sample should
be tested. Similar arguments hold true when considering a system composed of N
components.

Rule of thumb: To show with 95% accuracy that the percentage of damaged
elements in a production is at most $p = 10^{-x}$, a zero-defect sample of order $n = 3 \cdot 10^x$ is necessary.

Explanation: The rule says that we want to find an n such that

$$B_{n,p}(0) \leq 0.05, \tag{4.1}$$

where the binomial distribution is used on the left side and $p = 10^{-x}$. Computing
the left side for $n = 3 \cdot 10^x$ and $p = 10^{-x}$ yields

$$B_{n,10^{-x}}(0) = \binom{n}{0} \cdot p^0 \cdot (1-p)^n = (1-p)^n = \left(1 - 10^{-x}\right)^{3 \cdot 10^x} \approx e^{-3} \approx 0.0498. \tag{4.2}$$

For systems with strict regulations, the numbers derived from the rule of thumb can soon lead to unrealistically big samples.

If the method of examining samples fails, theoretical system analysis method becomes very important. Hence, it becomes the only method that can guarantee the dependability of a production entity or the safety of a system.

4.5 Inductive and Deductive System Analysis Methods

There are two different concepts for system analysis: inductive and deductive methods.

Inductive system analysis methods analyze in which way components can malfunction and the consequences on the whole system. That is, an inductive approach means reasoning from an individual consideration to a general conclusion.

Deductive methods analyze how a potential failure of the whole system or a critical system fault, for example, the maximum credible accident, can occur and what has to happen for this to occur. That is, a deductive approach means reasoning from a general consideration to special conclusions.

Figure 4.2 visualizes the differences of inductive (top sequence) and deductive (bottom sequence) methods. The big square represents the system. The smaller squares are the components and dashed squares are malfunctioning components.

For more details about the classification of methods see Table 4.1. The listed methods will be explained in Chap. 5. Fault tree analysis (FTA) will be regarded as an example for a deductive system analysis method in Chap. 6 and failure modes and effects analysis (FMEA) as an example for an inductive system analysis method in Chap. 7.

It is also possible to combine inductive and deductive methods. This is, for example, done in the "Bouncing Failure Analysis" (BFA) (Bluvband et al. 2005).

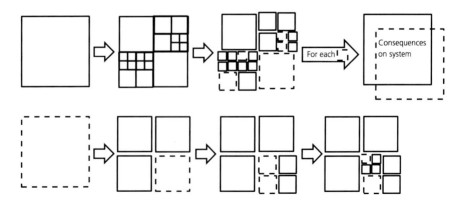

Fig. 4.2 Comparison of an inductive (top) and a deductive (bottom) approach

Table 4.1 Inductive and deductive system analysis methods

Method	Inductive	Deductive
Description	Analysis in which way components can malfunction and the consequences on the whole system	Analysis how the whole system can fail and what has to happen for this to occur
Examples	"Parts Count" (PC) "Failure Modes and Effects Analysis" (FMEA) "Hazard Analysis" (HA) with respect to hazard types "Double Failure Matrix " (DFM) "Fault Hazard Analysis " (FHA) "Event Tree Analysis" (ETA) (Vesely et al. 1981)	"Fault Tree Analysis" (FTA) "Dynamic Fault Tree Analysis" (DFTA)
Related question	What happens if…?	How could it happen that…?
Advantage	Often more effective in finding most likely failures of the whole system (if single-point failures dominate)	More effective in finding elementary undetected malfunction sources
	Considers the influence of every part	Considers the combined influences of parts
Disadvantage	Less effective in finding malfunction sources for a given system fault	Difficult to detect all relevant combinations of failures leading to overall system failure (especially without profound system knowledge and systematic FTA-building method)
	Does only consider single faults or up to double faults	All relevant possible (top level) overall system faults must be known before the analysis

4.6 Forms of Documentation

Two forms of documentation of system analysis are widely used: tables and graphics.

Figure 4.3 shows a first overview of the columns in the worksheets of different classical inductive methods. The terms failure and fault are defined in Sect. 4.10.

Example FMEA is a method where results are documented in a table while the results of an FTA are documented in a graphic, see Chaps. 6 and 7, respectively.

A problem is to find a good balance between the formalities of these forms of documentation and the content. Focusing too much on formalities, i.e., "to mistake form for substance" (Vesely et al. 1981), yields less flexibility and not very deep results. Only focusing on the content, however, makes it hard for other persons to understand and process the results.

Parts count approach

| component designation | failure probability (e.g. failure rate, unavailability) |

Failure Mode and Effect Analysis (FMEA)

| component designation | failure probability | component failure modes |

percentage of total failures attributable to each mode | effect on overall system (e.g.

MIL STD 882: "catastrophic", critical", "marginal", "negligible") |

Failure Mode Effect and Criticality Analysis (FMECA)

FMEA | detailed description of potential effects of faults | existing or projected

compensations or controls | summary |

Preliminary Hazard Analysis (PHA)

| component | hazard mode |compensation and control | findings and remarks |

Fault Hazard Analysis (FHA)

| component identification | failure probability |failure modes | percent failures by

mode | effect of failure | identification of critical upstream component | factors for

secondary failures| remarks |

Fig. 4.3 Overview of the columns of inductive FMEA like methods according to (Vesely et al. 1981)

4.7 Failure Space and Success Space

Besides choosing the method and the form of documentation, there is another choice that has to be made. Analysis can take place in the failure space or in the success space. In the *failure space*, the possibilities of failures are analyzed, for example, the event that an airplane does not take off. In the *success space*, the opposite event would be analyzed, that is, that the airplane does take off.

Theoretically, both approaches are of same value but analyzing the failure space has turned out to often be the better choice in practical applications. One reason is that it is often easier to define a failure than a success (Vesely et al. 2002). Another reason is that there usually are more ways to be successful than to fail. Hence, working in the failure space often is more efficient.

For some methods, it is possible to transfer the results from the failure space into the success space. For example, in classical fault tree analysis, one can compute the logical compliment of the fault tree which then lies in the success space. However, in general, this does not yield the same tree as an analysis in the success space.

Use of success space is heavily advocated by Safety 2.0 and resilience engineering approaches, see, e.g., (Woods et al. 2017).

4.8 Overview Diagram

The necessary steps one has to go through before starting the actual analysis and the choices one has to make are summarized in Fig. 4.4.

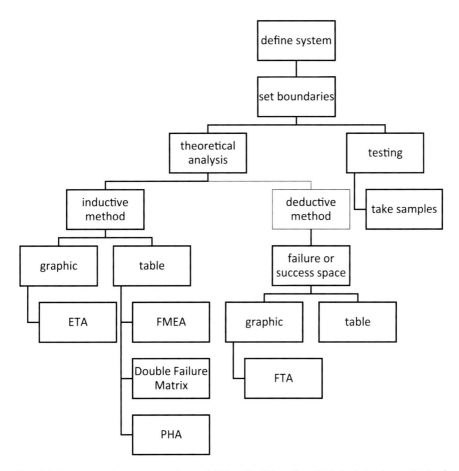

Fig. 4.4 Overview of system analysis possibilities. We did not list all tabular inductive methods of Table 4.1

4.9   Black Swans

Two different types of risks are distinguished. Risks of which it is known that they exist are called *known unknowns*. Risks of which it is not known that they exist are called *unknown unknowns* or *black swans* (Taleb 2007).

4.10   Failure and Fault

It is important to distinguish between failures and faults (Vesely et al. 1981) since they play different roles in the system analysis process. A *failure* is a malfunction within a component while a *fault* can also occur because the component is addressed in the wrong way.

Example A defect in the external control mechanism of a machine could cause the engine to start in the wrong moment. This would not be a failure within the component "engine" but a fault. If a part of the engine was broken, this would be a failure of the component "engine."

In general, every failure is a fault but not every fault is a failure.

4.11   Types of Failures

One distinguishes between three different types of failures, "primary, secondary and command failures" (Vesely et al. 1981).

Primary failures are failures where a component or system fails under conditions for which it was developed.

Secondary failures are failures where a component or system fails under conditions for which it was not developed.

Command failures are failures that are caused by a wrong command or a command at a wrong time. They also comprise failures caused by wrong handling of a component or the whole system.

Example Suppose a gas tank was developed for a maximal pressure p_0. A primary failure would be if the tank leaks at a pressure $p \leq p_0$. A secondary failure would be if it leaks at a pressure $p > p_0$. A command failure would be if a person accidentally opens the emergency valve.

These failure types are used in a systematic way when constructing fault trees, see Sect. 6.5.

4.12 Safety and Reliability

Safety and reliability are two aims of system analysis and often contradict each other. We have already seen a general definition in Sect. 3.11. In that section, safety was defined as the absence of intolerable risks (IEC 61508 S+ 2010). Reliability meant that components are always sufficiently active when they are supposed to be.

In terms of systems, this can be expressed in the following way:

Reliability means that reliability (comfort) functions of the systems are performed reliably.

Safety, on the other hand, means that the safety functions of the system are performed reliably.

This can be achieved if the system is reliable, and active safety functions further reduce risks if necessary. The safety functions must be sufficiently reliable. So, safety for typical systems (where safety functions are added to reliability functions or mixed with them) considers both, reliability and safety functions.

Example A machine that cannot be started at all is safe but not reliable. A device that starts when the power button is pressed but also when being dropped on the ground is reliable but not safe. Only a device that starts every time the button is pressed and only when it is pressed is safe and reliable.

The following example shows that it is not always easy to set clear boundaries between safety and reliability.

Example The reliability function of an air bag is to function in case of a crash event. The safety function of an airbag is to avoid unintended functioning during driving if no crash occurs. When determining the necessary reliability of the safety function of the airbag, the reliability of the failure rate of the reliability/comfort function has to be considered. It might even not be necessary to use an additional safety function and corresponding safety-related systems. Typical levels are SIL 1 for the reliability function and SIL 3 for the safety function which can be derived from analyzing the norms ISO 26262 and IEC 61508 in combination. In the latter case, the same subsystem of the airbag has to fulfill the requirements for the reliability/comfort function and the safety function. Obviously, in this case, making the airbag more reliable might make it less safe.

Supplementary protection measures (machinery emergency stops) can also be considered as safety functions (Zehetner et al. 2018). However, in this case, there are persons in the loop that operate as human sensors within a generalized sensor-logic-actor chain that implements the safety function. Hence, the reliability of such functions is often not considered as very high and machine safety typically does not rely on it.

4.13   Redundancies

Redundancies are a common method to improve the safety and/or reliability of a system. Redundancies are additional parts in a system that are not needed if the system works without any failures.

When regarding a system in terms of redundancies, one has to keep in mind that failures often have influences on several components. Even if one has applied additional parts, there might be *non-independent failures* in the redundant parts. A special case of non-independent failures are so-called *common-cause failures*. These are failures in two different components or subsystems with common origin (root cause).

Example To ensure that a pump is supplied with enough water, one can install two pipes that transport water independently of each other. The pump only fails if both pipes fail. Unfortunately, the probability of a failure of the pump is not the product of the probability of the failure of each individual pipe because non-independent is possible. For example, the probability of the remaining pipe to fail can increase if the other pipe fails because it is used more intensely. A common-cause failure would, for example, be a fire that destroys both pipes.

The example shows that redundancies often only improve the safety or reliability of a system for certain types of failures and that one has to be careful when calculating probabilities for a redundant system.

Due to the oppositeness of safety and reliability, see Sect. 4.12, redundancies that make a system safer often make it less reliable and vice versa. To increase the safety and reliability of a system by installing redundancies, it is, hence, necessary to have separate redundancies for the safety and the reliability (sub)systems.

Example (compare to the example in Sect. 4.12)**:** A further sensor of the airbag system increases its reliability but decreases safety.

Example A sensor detecting if only a small child is sitting in front of the airbag makes the airbag safer but decreases its reliability because it cannot be started if the child detection has a failure.

4.14   Active and Passive Components

It is useful to distinguish between active and passive components in the system analysis because *passive components*, for example, wires or screws, are often more reliable than *active components*, for example, pumps or switches (Vesely et al. 1981).

4.15 Standby

The term cold standby is used if a redundant component is not used in normal system operation, but only in case of major system degradation of failure. In case of hot standby, the component is used in normal operating condition but typically there is at least some or even significant reduction of capacity. Warm standby means that the component is in an intermediate status of employment during standard system operation.

4.16 Optimization of Resources

Another aspect of system analysis is the identification of unnecessary measures.

Example Two components that both are not allowed to fail have probabilities of failures of 10^{-3} and 10^{-7}, respectively. It would not be useful to make the already much more reliable component even more reliable although this might be easier than to improve the other component.

System analysis detects and avoids these unnecessary measures and helps to focus on essential improvement.

4.17 Combination of Failures

Several system analysis methods, for example, fault tree analysis or the double matrix method, include the possibility to analyze combinations of failures. This is important for large and complex systems because minor failures might cause a lot more damage if they happen simultaneously or in a certain order. An example where this happened is the development of the American space shuttle (Vesely et al. 2002).

4.18 FTA and FMEA in the Context of Other Methods

Chapter 4 introduced definitions used within system analysis methods. The most important terms are roughly ordered according to their use within analysis processes. An overview of different system analysis methods has also been given in Fig. 4.4.

Following as sample applications of the scheme of Fig 4.4. The paths

define system—set boundaries–theoretical analysis–failure space–deductive method–graphic–FTA

and

define system–set boundaries–theoretical analysis–failure space–inductive method–table–
FMEA

leads to the two most often used system analysis methods that have already been
mentioned in this chapter but will be introduced in detail in Chaps. 6 and 7 namely,
fault tree analysis (FTA) and failure modes and effects analysis (FMEA).

4.19   Questions

(1)   Define system analysis and system simulation. Name typical steps. How are
      they related to system definition?
(2)   Which type of boundaries determine the resolution of the analysis?
(3)   How many zero-defect samples are approximately necessary to show with
      95% level of confidence that the percentage of damaged elements is at most
      $p = 0.001$?
(4)   How is a system classically defined?
(5)   Why is the classical systems definition difficult to fulfill for modern systems?
(6)   Give an example for each of the four system boundary types.
(7)   What is the difference between the approach of deductive and inductive system
      analysis methods?
(8)   How can a fault tree be transferred from the failure space to the success space?
(9)   What is the difference between failure and fault?
(10) Several mechanical components are slightly out of their tolerance values
      combining in an unfavorable way.
(11) Distinguish three generic failure types of systems that can be expected in
      systems!

4.20   Answers

(1)   See Sects. 3.17, 3.18 and 4.3.
(2)   See Sect. 4.3.
(3) $p = 0.001 = 10^{-3}$ asks for $n = 3 \cdot 10^3 = 3,000$ samples, compare Sect. 4.4.
(4)   See Sect. 4.2.
(5)   See Sect. 4.3 for the definitions. Some examples are
      External: If the analyzed system is the camera in a smartphone, the sim card
      and its holder are not considered in the analysis.
      Internal: The engine is analyzed as one component. The different parts of the
      engine are not regarded.
      Temporal: A smartphone is expected to work for 3 years.
      Probabilistic: The start of the engine fails with a frequency of 10^{-4}. The faults
      leading to this are not further resolved.

(6) For instance, in case of a high-altitude wind energy system using soft kites and tethers, the external boundary could include the output interfaces of the generator and interfaces to the overall system steering. The internal boundaries could be defined by the components that are not further changed during production regarding their physical-engineering properties, e.g., hardware components. The time boundary is the overall expected in-service time. Probability boundaries could be the lowest failure probability of a component that is still considered, e.g., $10^{-9}\ h^{-1}$.

(7) See Table 4.1.

(8) See Sect. 4.7.

(9) See Sect. 4.10.

(10) See Sects. 4.10 and 4.11.

(11) See Sects. 4.10 and 4.11.

References

Bedford, T. and R. Cooke (2001). Probabilistic Risk Analysis: Foundations and Methods, Cambridge University Press.

Bluvband, Z., R. Polak and P. Grabov (2005). "Bouncing Failure Analysis (BFA): The Unified FTA-FMEA Methodology." Annual Reliability and Maintainability Symposium, 2005. Proceedings.: 463–467.

Echterling, N., A. Thein and I. Häring (2009). Fehlerbaumanalyse (FTA) - Zwischenbericht.

ISA4.0 (2020): BMBF Verbundprojekt: Intelligentes Sensorsystem zur autonomen Überwachung von Produktionsanlagen in der Industrie 4.0 - ISA4.0, Joint research project: Intelligent sensor system for the autonomous monitoring of production systems in Industry 4.0, ISA4.0, 2020–2023.

IEC 61508 S+ (2010): Functional Safety of Electrical/Electronic/Programmable Electronic Safety related Systems Ed. 2. Geneva: International Electrotechnical Commission.

Maluf, D. A. and O. Gawadiak (2006). "On Space Exploration and Human Error: A paper on reliability and safety."

Müller, D. H. and T. Tietjen (2003). FMEA-Praxis, Hanser.

Taleb, N. N. (2007). The Black Swan: The Impact of the Highly Improbable.

ULTRA-FLUK (2020): Ultra-Breitband Lokalisierung und Kollisionsvermeidung von Flug-drachen, Ultra-broadband localization and collision avoidance of kites, BMBF KMU-Innovativ Forschungsprojekt, German BMBF research project, 2020–2023.

Vesely, W. E., F. F. Goldberg, N. H. Roberts and D. F. Haasl (1981). Fault Tree Handbook. Washington, D.C., Systems and Reliability Research, Office of Nuclear Regulatroy Research, U.S. Nuclear Regulatory Comission.

Vesely, W., M. Stamatelatos, J. Duagan, J. Fragola, J. Minarick and J. Railsback (2002). Fault Tree Handbook with Aeorspace Applications. Washington, DC 20546, NASA Office of Safety and Mission Assurance, NASA Headquaters.

Woods, David D.; Leveson, Nancy; Hollnagel, Erik (2017): Resilience Engineering. Concepts and Precepts. 1st ed. Boca Raton: CRC Press.

Zehetner, Julian; Weber, Ulrich; Häring, Ivo; Riedel, Werner (2018): Towards a systematic evaluation of supplementary protective measures and their quantification for use in functional safety. In S. Haugen, A. Barros, C. van Gulijk, T. Kongsvik, J. E. Vinnene (Eds.): Safety and Reliability - Safe Societies in a Changing World. Safety and Reliability – Safe Societies in a Changing World, Proceedings of the 28-th European Safety and Reliability Conference (ESREL), Trondheim, Norway, 17–21 June 2018: CRC Press, pp. 2497–2505. Available online at https://www.taylorfrancis.com/books/9781351174657, checked on 9/23/2020.

Chapter 5
Introduction to System Analysis Methods

5.1 Overview

This chapter gives an overview of classical system analysis methods. Mostly, a representative example is given for each of the methods to get a general idea what is hidden behind the name. It is not intended to be sufficient to actually use the method. However, it aids to support the selection of the correct type of method by listing the main purposes of the methods.

The chapter takes up the methods of Table 4.1 which was introduced in Sect. 4.5. The methods listed in this table will be briefly introduced in this chapter. The main categorization available in terms of graphical versus tabular, inductive versus deductive, and qualitative versus quantitative is refined in this chapter by considering in addition typical implementation examples, phases of developments and life cycles where the methods are used as well as typical application examples.

The first method explained in more detail is the fault tree analysis in Chap. 6.

5.2 Parts Count Approach

In the *Parts Count* approach, every component failure is assumed to lead to a critical system failure. It is a simple version of a failure modes and effects analysis (FMEA). A very conservative upper bound for the critical failure rate is obtained. It also takes common-cause and multiple failure modes into account, see Fig. 5.1.

"Suppose the probability of failure of amplifier A is 1×10^{-3} and the probability of failure of amplifier B is 1×10^{-3}, i.e., $f_A = 1 \times 10^{-3}$ and $f_B = 1 \times 10^{-3}$. Because the parallel configuration implies that system failure would occur only if both amplifiers fail, and assuming independence of the two amplifiers, the probability of system failure is $1 \times 10^{-3} \times 1 \times 10^{-3} = 1 \times 10^{-6}$. By the parts count method, the component probabilities are simply summed and hence the 'parts count system

© The Author(s), under exclusive license to Springer Nature Singapore Pte Ltd. 2021
I. Häring, *Technical Safety, Reliability and Resilience*,
https://doi.org/10.1007/978-981-33-4272-9_5

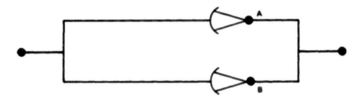

Fig. 5.1 A system of two amplifiers in parallel (Vesely et al. 1981, Figure II-1)

failure probability' is $1 \times 10^{-3} + 1 \times 10^{-3} = 2 \times 10^{-3}$ which is considerably higher than 1×10^{-6}" (Vesely et al. 1981).

5.3 Failure Modes and Effects Analysis (FMEA)

The failure modes and effects analysis (FMEA) considers for every component the (overall) failure probability, the modes, and their percentage that contribute to the failure probability and the effects on the total system (Vesely et al. 1981). The effects on system level are classified as critical and non-critical.

Table 5.1 shows an example of an FMEA for the system of Fig. 5.1.

Remark Here, the parts count critical system failure for the system of Fig. 5.1 according to the FMEA table given in Table 5.1 is 2×10^{-3} by summing up the values in column 2 and the FMEA critical system failure reads 2×10^{-4} by summing up the values in the "critical effects" column in column 5.

Birolini defines failure fractions (percentages of failure modes) for important electronic component types, see Table 5.2.

Table 5.1 Example of an FMEA for the system of Fig. 5.1 (Vesely et al. 1981, Table II-1)

1	2	3	4	5	
Component	Failure probability	Failure mode	% Failures by mode	Effects	
				Critical	Non-critical
A	1×10^{-3}	Open	90		X
		Short	5	X (5×10^{-5})	
		Other	5	X (5×10^{-5})	
B	1×10^{-3}	Open	90		X
		Short	5	X (5×10^{-5})	
		Other	5	X (5×10^{-5})	

Table 5.2 Example for failure modes of electronic components; S = short circuit; O = open circuit/break; D = drift; F = failure, after Birolini (1994)

Component	S (%)	O (%)	D (%)	F (%)
Digital bipolar integrated circuits (IC)	30	30	10	30
Digital MOS-ICs	20	10	30	40
Linear ICs	30	10	10	50
Bipolar transistors	70	20	10	–
Field-effect transistors (FET)	80	10	10	–
Diode	70	30	–	–
Zener diode	60	30	10	–
High-frequency (HF) diode	80	20	–	–
Thyristors	20	20	60	–
Optoelectronic components	10	50	40	–
Resistor	≈0	90	10	–

Table 5.3 shows the FMEA of the withdrawn DIN 25488 (DIN 25448 1990). Column 5 focuses on the detection of faults and column 6 on countermeasures. It is interesting that there is no column for the estimation of a failure rate.

FMEA will be treated in more detail in Chap. 7, including more recent developments.

5.4 Failure Modes, Effects and Criticality Analysis (FMECA)

The failure modes, effects and criticality analysis (FMECA) is very similar to the FMEA. The criticality of the failure is analyzed in greater detail. Assurances, controls, and compensations are considered that limit the probability of the critical system failure (Vesely et al. 1981). A minimum set of columns includes fault identification, potential effects of the fault, existing or projected compensation, and control and summary of findings.

A typical FMECA form is shown in Fig. 5.2. It is based on the withdrawn DIN 25488 (DIN 25448 1990). Additional material is added from the electro and automotive industry domain (Leitch 1995; Witter 1995; Birolini 1997).

The risk priority number (RPN) is defined in withdrawn DIN 25488 (DIN 25448 1990). It is the product of the severity (S), the occurrence rating (0), and the detection rating (D):

$$RPN = S \ O \ D. \tag{5.1}$$

It is important to note that the detection rating is actually a non-detection rate.

Table 5.3 Classical worksheet example preprinted tabular form for FMEA, taken from the withdrawn DIN 25448 1990)

Failure mode and effect analysis (FMEA)							Page 1
System: Distribution system for pressurized air	Subsystem: Pressure vessel						
Initial state	Initial state					Reference documentation	
Undisturbed operation, operation as specified, pressurized mode	Maximum allowable working pressure (MAWP)[a] = 1.5 MPa Maximum allowable operating pressure (MAOP) = 1.0 MPa		Room temperature 10 to 30 °C Air humidity ≤ 80% Dust-free atmosphere			Drawings Specifications	
1	2	3	4	5	6	7	8
No.	Function	Failure mode	Damage symptoms, possible causes	Damage detection	Implemented countermeasures	Damage effects on system/consequences for system and on its surroundings if applicable	Effect, comments
1.1	Storage of pressurized air	Small leakage	Vessel connection leaking	Frequency of operation/on–off operations of compressor increased, detection of modified/increased compressor noise	Compressor feeds pressure vessel	Pressure decrease is compensated by operation of compressor	Maintenance and repair case, shutdown of system in case of repair
1.2	Storage of pressurized air	Large leakage	Crack of a weld seam	Pressure display and pressure surveillance, inspection	None	Fast pressure decrease	System failure

(continued)

Table 5.3 (continued)

Failure mode and effect analysis (FMECA)					Page 1
1.3	Storage of pressurized air	Fracture	Material defect or damage effects due to transport of pressure vessel	Pressure display and pressure surveillance, inspection	None	Fast pressure decrease, damage effects on facilities in the vicinity of the pressure vessel	System failure, state of high risk

[a]Definition see DIN 2401 Part 1

Company	FMEA type: System/Idea, Design/Construction, Production/Process				Model/System			Model/System number	
Date	Confirmed by	Date		Developed by	Date		Revised by		Date
System state considered									

System description	Failure Analysis					Risk Assessment				Counter activities		Improvement measure assessment			
						Column number									
1	2	3	4	5	6	7	8	9	10	11	12	13	14	15	16	17	18	19	20

1	2	3	4	5	6	7	8	9	10	11	12	13	14	15	16	17	18	19	20
System, Component number	Element	Function	Failure mode	Failure cause	Failure consequence	Failure mitigation	Failure detection	Severity of consequence S	Failure rate λ	Occurrence rating O	(Non-) Detection rating D	Risk Priority Number RPZ	Recommended counter measure	Responsible person	Implemented counter measure	Severity	Occurrence	Detection	RPZ

Fig. 5.2 Worksheet implementing an FMECA, based on DIN 25448 (1990) with additional material from Leitch (1995), Witter (1995), Birolini (1997)

The sizes S, O, and D are integer numbers between 1 and 10. For instance, S = 1 means the severity of the consequences is negligible and O = 1 the occurrence is very low.

For quantitative considerations, a correspondence table is necessary that relates the semi-quantitative numbers that only lead to an ordering to real numbers. Examples for S are measures for consequences, e.g., fatalities per failure, per hour, per year, or individual risks for the user. Examples for O are failure rates per hour or on demand. An example for D is percentages of detection.

Often quite arbitrary one requires that

$$RPN \leq RPN_{critical} \tag{5.2}$$

or, individually for each factor,

$$S \leq S_{critical},$$
$$O \leq O_{critical},$$
$$D \leq D_{critical}. \tag{5.3}$$

Also, products of two factors can be used. FMECA will be treated in more detail in Chap. 7.

5.5 Fault Tree Analysis (FTA)

Fault tree analysis (FTA) uses a tree-like structure and Boolean algebra to determine the probability of one undesired event, the so-called top event. It is also determined which combination of basic events leads to this top event. It can be used to find out which components are the most efficient to improve. An example for such a tree is given in Fig. 5.3.

An event above an "Or" gate occurs if at least one of the events below the gate occurs. An event above an "And" gate occurs if all events below the gate occur. For example:

– A student fails the exam if it is the student's fault or if it is the teacher's fault.
– It is a student's fault if the student made too many mistakes and if the exam was fair.

FTA will be explained in more detail in Chap. 6.

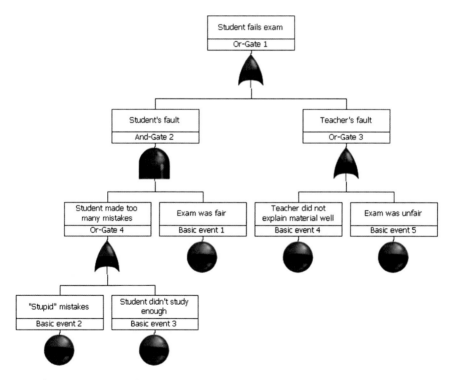

Fig. 5.3 A simple example fault tree, created with the software tool (Relex 2011). It is assumed that the cause is not a combination of faults by the student and teacher, that is, we assume a simplified "black and white" world. Otherwise, the tree would be considerably larger

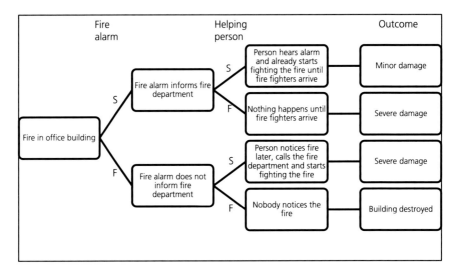

Fig. 5.4 A simple event tree. S = success, F = failure

5.6 Event Tree Analysis (ETA)

Event tree analysis (ETA) uses a structure which resembles a lying tree where the root of the tree is at the left-hand side of the worksheet. The root of the tree represents an initiating event. Moving to the right the possible outcomes of the event are described, see Fig. 5.4. On the way, so-called chance nodes distinguish between possible following events, for example, by whether the fire alarm does or does not go off.

"Note that each chance node must have branches that correspond to a set of outcomes that are mutually exclusive (i.e. only one of the outcomes can happen) and collectively exhaustive (i.e. no other possibilities exist, one of the specified outcomes has to occur). The probabilities at each chance node must add up to 1." (ReliaSoft Corporation 2013)

The systematic combination of FTA and ETA is often called bow-tie analysis (Ruijter and Guldenmund 2016).

5.7 Hazard Analysis (HA)

Hazard analysis (HA) focusses on hazardous components or events and their resulting risks to persons and objects. This is in contrast to Parts Count, FMEA, and FTA which all focus on all faults of components and their consequences for the system. The results of FMEA and FTA are used in the hazard analysis.

Hazard analysis is an inductive system analysis method where results are documented in a table. Typical columns read

- Component/subsystem and hazardous modes,
- Possible effects,
- Countermeasures and verification, and
- Findings and remarks.

Hazard analysis will be explained in more detail in Chap. 8.

5.8 FHA

The *fault hazard analysis* (FHA) starts to take more than only a single component or mode into account. It looks in more detail on causes and consequences of faults: it looks upstream and downstream.

"A typical FHA form uses several columns as follows:

Column (1)-Component identification.
Column (2)-Failure probability.
Column (3)-Failure modes (identify all possible modes).
Column (4)-Percent failures by mode.
Column (5)-Effect of failure (traced up to some relevant interface).
Column (6)-Identification of upstream component that could command or initiate the fault in question.
Column (7)-Factors that could cause secondary failures (including threshold levels). This column should contain a listing of those operational and environmental variables to which the component is sensitive.
Column (8)-Remarks" (Vesely et al. 1981).

5.9 DFM

The *double failure matrix* (DFM) determines the effects of two faults in a systematic way. The possible outcome of faults is assessed using a verbal qualitative description of system effects, see, for example, Tables 5.4 and 5.5.

Table 5.6 shows the effects of the double failure modes of Fig. 5.5.

Table 5.4 Fault categories and corresponding system effects (Vesely et al. 1981, Table II-2)

Fault category	Effect on system
I	Negligible
II	Marginal
III	Critical
IV	Catastrophic

Table 5.5 Fault categories of the NERVA (Nuclear Engine for Rocket Vehicle Application) Project (Vesely et al. 1981, Table II-3)

Fault category	Effect on system
I	Negligible
IIA	A second fault event causes a transition into Category III (critical)
IIB	A second fault event causes a transition into Category IV (catastrophic)
IIC	A system safety problem whose effect depends upon the situation (e.g., the failure of all backup onsite power sources, which is no problem as long as primary, offsite power service remains on)
III	A critical failure and mission must be aborted
IV	A catastrophic failure

5.10   Questions

(1)   Suppose the independent components A and B with known failure probabilities are connected in series (see Fig. 5.6) and the Parts Count method is used to compute the total probability of failure.

In this case, the Parts Count method.

(a)   underestimates the failure,
(b)   overestimates the failure, and
(c)   computes the failure correctly.

(2)   Take the following line from Table 5.2:

Component	Short	Open	Drift	Failure
Bipolar transistors	70%	20%	10%	-

Assume that the transistor in your project has the failure rate $7.4 \times 10^{-4}\ h^{-1}$ (base failure rate for bipolar transistors with low frequency from the standard MIL-HDBK 217F).
Take the following extract of an FMEA table:

Component	Failure probability	Failure mode	% Failures by mode	Failure Probability by mode

How would you fill this table for your bipolar transistor?

Table 5.6 Fuel system double failure matrix (Vesely et al. 1981, Table II-4)

		BVA		CVA		BVB		CVB	
		Open	Closed	Open	Closed	Open	Closed	Open	Closed
BVA	Open	(One Way) IIA	✕	III	IIB	IIA	IIA or IIB	IIA	IIA
	Closed	✕	(Two Ways) IIB	IIB	IIB	IIA or IIB	IV	IIA or IIB	IV
CVA	Open	III	IIB	(One Way) IIA	✕	IIA	IIA or IIB	IIA	IIA or IIB
	Closed	IIB	IIB	✕	(Two Ways) IIB	IIA	IV	IIA or IIB	IV
BVB	Open	IIA	IIA or IIB	IIA	IIA or IIB	(One Way) IIA	✕	III	IIB
	Closed	IIA or IIB	IV	IIA or IIB	IV	✕	(Two Ways) IIB	IIB	IIB
CVB	Open	IIA	IIA or IIB	IIA	IIA or IIB	III	IIB	(One Way) IIA	✕
	Closed	IIA or IIB	IV	IIA or IIB	IV	IIB	IIB	✕	(Two Ways) IIB

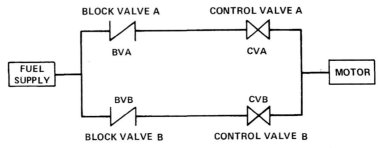

Figure II-2. Fuel System Schematic

Fig. 5.5 Fuel system schematic (Vesely et al. 1981, Figure II-2)

Fig. 5.6 Two components
connected in series

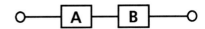

(3) Reconsider the example fault tree from Fig. 5.3.
     Find all basic events (marked with a purple circle) and all combinations of basic
     events that lead to the event "Student fails exam."
(4) Set up an event tree for the event "Student copies the neighbor's test" considering
     whether or not the neighbor knew the correct answers and whether or not the
     teacher noticed.
     While you create the tree, notice how much more complex it would be if you
     also considered that the student already knew enough answers to pass by himself
     and copying would only help him to achieve a better grade. It gives you a small
     impression how fast such kind of tree grows in real-life situations.
(5) Which methods would you recommend to be conducted before FTA?

5.11 Answers

(1) Parts count estimate $P(S) = P(A \cup B) \approx P(A) + P(B)$, which is more or
     equal than $P(A) + P(B) - P(A \cap B) = P(A) + P(B) - P(A)P(B)$. Hence,
     answer (b) is correct.
(2) The table is as follows:

Component	Failure probability	Failure mode	%Failures by mode	Failure probability by mode
Bipolar transistor, component number	0.00074	Short	0.7	0.000518
		Open	0.2	0.000148
		Drift	0.1	0.000074
		Failure	0	0

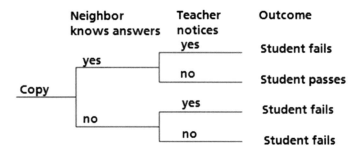

Fig. 5.7 Example event tree

(3) The basic events 4 and 5 lead to the top event by themselves because there are only "Or" gates above them. Basic event 1 only leads to the top event if also 2 or 3 occurs. Hence, the answer is (1, 2), (1, 3), 4, 5.

(4) The answer is given in Fig. 5.7.

(5) FMEA, to know the system better and all single faults. DFM to know all combinations of faults and to start considering combinations of faults. Generalizations of FMEA (e.g., with column which additional failure leads to system failure) and DFM (e.g., considering triple failures, etc.).

References

Birolini, A. (1994). Quality and Reliability of Technical Systems. Berlin, Heidelberg, Springer-Verlag.

Birolini, A. (1997). Zuverlässigkeit von Geräten und Systemen, Springer-Verlag.

DIN 25448 (1990). Ausfalleffektanalyse (Fehler-Möglichkeits- und Einfluss-Analyse); Failure mode and effect analysis (FMEA). Berlin, DIN Deutsches Institut für Normung e.V.; Beuth Verlag.

Leitch, R. D. (1995). Reliability Analysis for Engineers, An Introduction, Oxford Science Publications.

Relex (2011). Relex reliability Studio Getting Started Guide.

ReliaSoft Corporation. (2013). "Modeling Event Trees in RENO." Retrieved 2013-10-28, from http://www.weibull.com/hotwire/issue63/hottopics63.htm.

Ruijter, A. de; Guldenmund, F. (2016): The bowtiemethod: A review. In *Safety Science* 88, pp. 211–218. https://doi.org/10.1016/j.ssci.2016.03.001.

Vesely, W. E., F. F. Goldberg, N. H. Roberts and D. F. Haasl (1981). Fault Tree Handbook. Washington, D.C., Systems and Reliability Research, Office of Nuclear Regulatroy Research, U.S. Nuclear Regulatory Comission.

Witter, A. (1995). "Entwicklung eines Modells zur optimierten Nutzung des Wissens potentials einer Prozess-FMEA." Vortschrittsberichte VDI, Reihe 20, Nr.176.

Chapter 6
Fault Tree Analysis

6.1 Overview

Fault tree or root cause analysis (FTA) is the most often used graphical deductive system analysis. It is by now the working horse in all fields with higher demands regarding reliability, safety, and availability. It is increasingly also used in the domain of engineering resilience for assessing sufficient efficient response to expectable disruptions in terms of time needed to detect the disruption, to stabilize the system, and to repair or respond and recover.

The FTA approach is conducted in several steps including the identification of the goal of the analysis, of relevant top events to be considered, system analysis, FTA graphical construction, translation into Boolean algebra, determination of cut sets, probabilistic calculation of top events, importance analysis, and executive summary. Each step is illustrated and several computation examples are provided.

Major advantages of FTA include its targeted deductive nature that focuses on the system understanding and modeling used for the analysis; the option to conduct the analysis qualitatively and quantitatively; and the targeted evaluation options in terms of most critical root cause combinations (minimal cut sets), most critical cut sets, and root causes.

Further developments of FTA shortly discussed include its application to security issues in terms of attack trees, its application to the cybersecurity domain by typically assuming that all (or most) possible events also occur, and time-dependent FTA (TDFTA). TDFTA goes beyond considering system phase or use case-specific applications of FTA by considering time-dependent system states. Typically, such approaches are based on Markov or Petri models of systems. Another topic is to combine Monte Carlo approaches or fuzzy theory (fuzzification) to fault tree. Using distributions instead of failure probabilities allows to consider statistic uncertainties and deep uncertainties within FTA models of systems.

In summary, Chap. 1 introduces the deductive system analysis method FTA with focus on classical concepts, the process of successful application, and an overview on the many extension options. Section 6.2 gives a brief introduction to the method

© The Author(s), under exclusive license to Springer Nature Singapore Pte Ltd. 2021
I. Häring, *Technical Safety, Reliability and Resilience*,
https://doi.org/10.1007/978-981-33-4272-9_6

and mentions advantages, disadvantages, and application areas. Section 6.3 defines the most important terms and definitions used in FTA.

Three sections follow on the procedure of the analysis. Section 6.4 introduces the structure of FTA by dividing the analysis into eight well-defined steps with inputs and outputs. Section 6.5 explains three key concepts that should be applied during the analysis to simplify the search of possible events that cause failures, in particular, to search for immediate, necessary, and sufficient events for construction of the tree as well as by considering all types of possible failures. Section 6.6 lists more formal rules that should be followed when constructing a fault tree, e.g., that FTAs should not contain any gate-to-gate sequence.

Sections 6.7 to 6.11 deal with the mathematics of fault tree analysis. Rules of the Boolean algebra are introduced and applied in the computation of minimal cut sets. Here, the top-down and the bottom-up methods are explained with an example. After a short section about the computation of dual trees, Section 6.10 shows how the probability of the top event is computed using the inclusion/exclusion principle.

Section 6.11 introduces importance measures which help to identify the components of a fault tree in which improvement is most effective. The chapter ends with a brief outlook on ways to extend classical fault tree analysis in Sect. 6.12.

This chapter is a summary, translation, and partial paraphrasing of (Echterling et al. 2009) with some additional new examples and quotations.

6.2 Introduction to Fault Tree Analysis

Fault tree analysis is an example for a deductive system analysis method. It is a structured and standardized method to analyze an undesired event, the so-called top event. FTA determines not only the possible sources for this top event, but also the probability that the top event occurs. It is especially useful for big, complex, and dynamic systems (Blischke and Prabhakar Murthy 2003). It can be used to analyze and prevent potential problems and to analyze problems that have already occurred. It is a graphical form of system analysis that presents the logical path between the top event and all possible reasons for its occurrence.

Fault tree analysis was introduced more than 50 years ago and has been developed significantly over the years. It is still an active research area, trying to fulfill modern demands that come with more complex systems, compare Section 6.12.

Fault trees contain events and logical gates. These gates make it possible to transfer the graphical model into a mathematical model that is necessary to compute the probabilities of failures. The transformation will be explained later, see Section 6.7. Transferring the graphical model into a mathematical model yields three important aspects for the decision-maker (Kumamoto 2007).

- The probability of occurrence of the top event.
- The probability and significance of the combination of failure events that can cause the top event.

– The significance of individual components.

In most situations, a qualitative FTA already yields useful results at (sometimes significantly) reduced costs and its support in the justification of decisions should not be underestimated. However, if the decision-maker wishes a quantitative analysis, there is no reason not to add quantitative specifications to a qualitative FTA (Ericson 2005).

An advantage of FTA is that it is easy to understand and to apply and still gives useful insight into the system and discovers all possible reasons for the top event that is analyzed. The flexible graphical structure of FTA makes it possible to adapt the structure iteratively to the system design throughout the process.

Another positive time- and money-saving aspect of FTA is its focus on the top event. Events that do not influence the top event get eliminated when setting up the tree and are not part of the analysis. At the same time, this poses a risk. System components that are modified because of the FTA can cause failures which increase the probability of occurrence of a top event which belongs to the same system but is analyzed in a different fault tree.

Results of fault tree analysis can be used as follows.

– Verification of the system design's compatibility with existing safety requirements.
– Identification of problems that have occurred despite the existing safety requirements.
– Identification of failures with a common cause.
– Accompanying preventive measures which reduce or eliminate identified problems in a system.
– Introduction or modification of safety requirements for the next system design phase.

6.3   Definitions

To study fault trees, the following definitions are helpful.

6.3.1   Basic Event and Top Event

A *basic event* is an event that should not or cannot be analyzed further. Basic events have to be stochastically independent for a meaningful analysis, see Sect. 6.10.

The *top event* is the event for which the fault tree is set up.

Example A possible top event is "The computer does not start up to 5 seconds after pressing the starting button."

6.3.2 Cut Sets, Minimal Cut Sets, and Their Order

A *cut set* is a combination of basic events that causes the top event.

A *minimal cut set* is a cut set where every event in the combination is necessary in order for the combination to cause the top event. Minimal cut sets that contain one event are called minimal cut sets of *order* one, those with two events are of order two, etc.

Example Consider a chain with n links that is arranged in a circle, that is, a chain where the ends are also connected by a link. Define the top event as "The chain falls into two or more pieces." For this top event, two links have to break. So, there are $n(n-1)$ minimal cut sets (of order two) of the form "link a breaks"∩"link b breaks." The intersection of order three "link a breaks"∩"link b breaks" ∩"link c breaks" would be a cut set of order 3, but not minimal. The basic event "link a breaks" is not a cut set because it does not yield the top event. It would be a minimal cut set of order one for the top event "the chain breaks."

The union of all minimal cut sets contains all possible combination of events that lead to the top event. In the quantitative analysis, the probability of the union of all minimal cut sets hence yields the probability of occurrence of the top event. The minimal cut set with the highest probability of occurrence is called the *critical path*.

It is essential for the calculation of the probabilities that the basic events are independent. Otherwise, the fault tree computations in Section 6.10 underestimate the probability that the top event occurs.

6.3.3 Multiple Occurring Events and Branches

A *multiple occurring event* (MOE) is a basic event that occurs more than once in the fault tree.

A *multiple occurring branch* (MOB) is a branch of the fault tree that occurs more than once in the fault tree. All basic events in a MOB are by definition MOEs.

6.3.4 Exposure Time

Exposure time is the time a component is effectively exposed to a force. It has a big effect on the calculation of probabilities of failure.

6.4 Process of Fault Tree Analysis

When fault trees were first developed in the 1960s, they were seen as art (Vesely et al. 1981). It soon turned out that a set of commonly accepted rules eases the understanding, prevents misunderstandings, and allows different groups of persons to work in a team without complications.

We follow the structure of (Vesely et al. 2002) by dividing FTA into the following eight steps where steps 3 to 5 can be done simultaneously.

(1) **Identification of the goal of the FTA**: The goal of the analysis is phrased in terms of failures and faults to ascertain that the analysis fulfills the decision-maker's expectations.

(2) **Definition of the top event**: The top event should be phrased as precisely as possible, ideally using physical units (DIN IEC 61025 Entwurf 2004).

(3) **Definition of the scope of the analysis**: The outer boundaries of a system have to be set. If the system exists in different versions or has different modes, the used version or mode has to be specified.

(4) **Definition of the resolution**: It has to be decided how detailed the analysis should be. That is, the inner boundaries of the system have to be defined. It is also possible to include probabilistic or temporal boundaries, see Section 4.3. The right balance between a meaningful and precise analysis and cost efficiency has to be found.

(5) **Agreement on a set of rules**: To make the FTA accessible to other persons, one standard has to be chosen, for example, (DIN) IEC 61025 (DIN IEC 61025 2007), and applied consequently.

(6) **Construction of the fault tree**: The fault tree is constructed according to the chosen standard. The symbols that are used for events and gates vary in the different standards. Table 6.1 lists some basic and often used symbols. (DIN

Table 6.1 Often used symbols in fault trees

Symbol	Description
	The box contains text that describes the event.
	The circle stands for a basic event that cannot or should not be resolved further.
	The triangle informs that the fault tree is continued at a different place.
	"and gate: The events under this symbol cause this event if and only if all of them occur.
	"or" gate: The events under this symbol cause this event if and only if at least one of them occurs.

IEC 61025 2007) contains tables with the complete list of symbols for the standard IEC 61025 in its appendix.

(7) **Evaluation of the fault tree**: The qualitative evaluation yields the minimal cut sets (see Sect. 6.3.2). A quantitative evaluation additionally gives the probability of occurrence of the top event and, if desired, of the minimal cut sets. The failure tree can also be converted into a success tree which is its logical compliment (Vesely et al. 2002).

(8) **Interpretation and presentation of results**: The interpretation of the analysis and the presentation of results should not only show the probability that the top event occurs but also which minimal cut sets are mainly responsible and how the situation can be improved.

Table 6.1 lists the basic symbols used in FTA. The triangle is used for convenience.

6.5   Fundamental Concepts

The construction of a fault tree is an iterative process which starts at the top event and is repeated in every level of the tree down to the minimal cut sets and basic events. There are different concepts that can be applied on each level. Three important concepts are described in this section.

6.5.1   The I-N-S Concept

On each level one asks: What is immediate (I), necessary (N), and sufficient (S) to cause the event?

The *I-N-S concept* ensures that nothing is left out in the analysis by focusing on immediate rather than indirect reasons.

6.5.2   The SS-SC Concept

In the *SS-SC concept*, failures are divided into two categories: state of system (SS) and state of component (SC). A fault in an event box is classified by SS or SC depending on which type of failure causes it.

The advantage of the classification is that SC events always come out of an "or" gate while SS events can come out of an "or" or an "and" gate.

6.5.3 The P-S-C Concept

In the *P-S-C concept*, failures are divided into primary (P), secondary (S), and command (C) failures, see Sect. 4.11.

Command failures form the failure flow in a fault tree. If an analysis is complete, a comparison of the fault tree and the signal flow graph shows similarities.

6.6 Construction Rules

The following selection of important rules for the construction of a fault tree is from (Ericson 2005).

- Each event in the fault tree should have a name and be described by a text.
- The name should uniquely identify the event.
- The text should be meaningful and describe failures and faults. Text boxes should never be left empty.
- The failure or fault should be described precisely. If necessary, the text box should be extended. Abbreviating words is permitted, abbreviating ideas is not.
- All entrances of a gate should be completed before the analysis is continued on the next level.
- Gate-to-gate connections are not allowed.
- Miracles should not be assumed. It is not permitted to assume that failures compensate other failures.
- I-N-S, SS-SC, and P-S-C are concepts. Those concepts should be applied, but their abbreviations should not be written in the text boxes.

6.7 Mathematical Basics for the Computation of Fault Tree

For the computation of probabilities in fault trees, concepts from probability theory are used. For this, gates are modeled by mathematical symbols. The only two gates used in this chapter are the "or"- and the "and" gates. They can be expressed by union and intersection, respectively.

Example An "or" gate below event C with the events A and B as children (see Fig. 6.1) would be expressed by $C = A \cup B$.

For the standard methods used in this chapter, only standard stochastic formulas and the rules of the *Boolean algebra* are needed. They are summarized in Tables 6.2 and 6.3. Most of them can be understood with the diagram in Fig. 6.2.

For more complex methods, other mathematical concepts such as Markov chains are necessary. Those are not covered here. For an introduction to those more complex

Fig. 6.1 The event C occurs
if A or B occurs, created
with Relex 2011

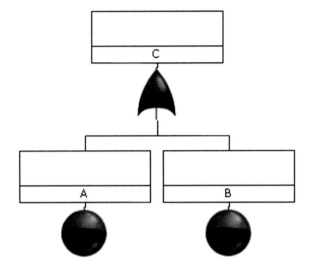

Table 6.2 Formulas to compute probabilities

Name	Formula
Probability of the complement	$P(\overline{A}) = 1 - P(A)$
Conditional probability	$P(B|A) := \frac{P(A \cap B)}{P(A)}$
Stochastically independent events	A and B are independent $\Leftrightarrow P(A \cap B) = P(A)P(B)$ $X_1, X_2, \ldots, X_n$ are independent $\Leftrightarrow P(X_1 \cap X_2 \cap \ldots \cap X_n) = P(X_1)P(X_2) \ldots P(X_n)$
Probability of the union of two events	$P(A \cup B) = P(A) + P(B) - P(A \cap B)$
Probability of the union of three events	$P(A \cup B \cup C) = P(A) + P(B) + P(C)$ $\qquad - (P(A \cap B) + P(A \cap C) + P(B \cap C))$ $\qquad + P(A \cap B \cap C)$
Probability of the union of two stochastically independent events	$P(A \cup B) = P(A) + P(B) - P(A)P(B)$ $\qquad = 1 - P(\overline{A})P(\overline{B})$
Probability of the union of n stochastically independent events	$P(X_1 \cup X_2 \cup \ldots \cup X_n) = 1 - \prod_{i=1}^{n} P(\overline{X}_i)$
Probability of the union of two events that cannot happen at the same time	$P(A \cup B) = P(A) + P(B)$
Probability of an exclusive union	$P(A \cup_{ex} B) = P(A \cup B) - P(A \cap B)$ $\qquad = P(A) + P(B) - 2P(A \cap B)$

Table 6.3 Boolean algebra

Name	Formula
Commutativity	$A \cap B = B \cap A$
	$A \cup B = B \cup A$
Associativity	$A \cap (B \cap C) = (A \cap B) \cap C$
	$A \cup (B \cup C) = (A \cup B) \cup C$
Distributivity	$A \cap (B \cup C) = (A \cap B) \cup (A \cap C)$
	$A \cup (B \cap C) = (A \cup B) \cap (A \cup C)$
Idempotence	$A \cap A = A$
	$A \cup A = A$
Absorption	$A \cap (A \cup B) = A$
	$A \cup (A \cap B) = A$
Complements	$A \cap \overline{A} = \varnothing$
	$A \cup \overline{A} = \Omega$
	$\overline{\overline{A}} = A$
Morgan's laws	$\overline{A \cap B} = \overline{A} \cup \overline{B}$
	$\overline{A \cup B} = \overline{A} \cap \overline{B}$
Extremal laws	$\varnothing \cap A = \varnothing$
	$\varnothing \cup A = A$
	$\Omega \cap A = A$
	$\Omega \cup A = \Omega$
	$\overline{\Omega} = \varnothing$
	$\overline{\varnothing} = \Omega$
Nameless	$A \cup (\overline{A} \cap B) = A \cup B$
	$\overline{A} \cap (A \cup \overline{B}) = \overline{A} \cap \overline{B}$

Fig. 6.2 Example Venn diagrams to illustrate Tables 6.2 and 6.3, e.g., the Boolean distributivity equations

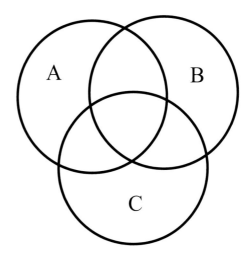

mathematical concepts and a more detailed and more elaborate introduction to probability theory, please refer to standard works on probability theory, for example (Klenke 2007).

6.8   Computation of Minimal Cut Sets

There are various methods to compute minimal cut sets which solve the fault tree. Two of these methods are the top-down method and the bottom-up method. Those two methods are explained in this section based on the example fault tree in Fig. 6.3.

The fault tree in Fig. 6.3 can be described by the following system of equations:

$$
\begin{aligned}
T &= E_1 \cap E_2, \\
E_1 &= A \cup E_3, \\
E_2 &= C \cup E_4, \\
E_3 &= B \cup C, \\
E_4 &= A \cap B.
\end{aligned}
\tag{6.1}
$$

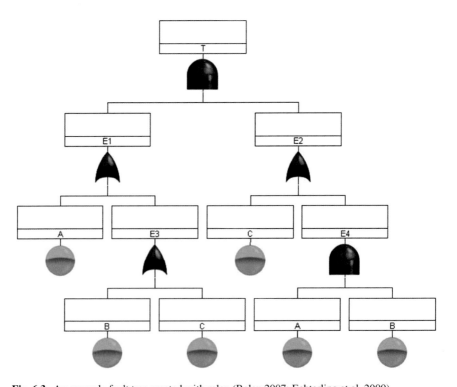

Fig. 6.3   An example fault tree created with relex (Relex 2007; Echterling et al. 2009)

Traditionally, $+$ is used instead of $\cup$ and $\cdot$ instead of $\cap$. With this notation (6.1) reads as

$$T = E_1 \cdot E_2,$$
$$E_1 = A + E_3,$$
$$E_2 = C + E_4,$$
$$E_3 = B + C,$$
$$E_4 = A \cdot B. \tag{6.2}$$

In the following, we use the notation from (6.2) but keep in mind that the rules of the Boolean algebra have to be applied. For example, $A \cdot A = A \cap A = A$ and not $A \cdot A = A^2$.

To avoid too many brackets, $A \cdot B + C \cdot D$ should be understood as $(A \cdot B) + (C \cdot D)$, as one would understand it for regular addition and multiplication.

6.8.1 Top-Down Method

In the *top-down method*, one starts with the top event and substitutes the equations in (6.2) from top to bottom into the first equation:

$$
\begin{aligned}
T &= E_1 \cdot E_2 \\
&= (A + E_3) \cdot (C + E_4) \\
&= (A + (B + C)) \cdot (C + (A \cdot B)).
\end{aligned}
\tag{6.3}
$$

Using the rules of the Boolean algebra, this can be written as

$$
\begin{aligned}
T &= (A + (B + C)) \cdot (C + (A \cdot B)) \\
&= ((A + B) + C) \cdot C + ((A + B) + C) \cdot (A \cdot B) \\
&= C + (A + B) \cdot (A \cdot B) + C \cdot (A \cdot B) \\
&= C + A \cdot B + C \cdot (A \cdot B) \\
&= C + A \cdot B,
\end{aligned}
\tag{6.4}
$$

where we have used associativity and distributivity in the first line, absorption and distributivity in the second, and absorption in the third and fourth lines.

In (6.4), it can be seen that the minimal cut sets are $\{C\}$ and $\{A, B\}$. Hence, an equivalent fault tree to the one in Fig. 6.3 is the one in Fig. 6.4.

Fig. 6.4 An equivalent fault
tree to the one in Fig. 6.3.
The tree was created with
relex (Relex 2007;
Echterling et al. 2009)

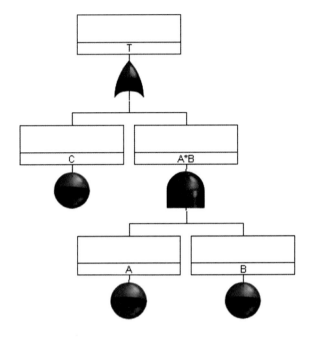

6.8.2 Bottom-Up Method

The *bottom-up method* calculates the minimal cut sets in a similar way to the top-down method. The only difference is that one starts by substituting the last equations in (6.2) into the equations above them and then continues toward the first equation. This yields

$$E_1 = A + (B + C),$$
$$E_2 = C + (A \cdot B), \tag{6.5}$$

and by substituting those into the first line

$$T = (A + (B + C)) \cdot (C + (A \cdot B)). \tag{6.6}$$

The calculation in (6.4) can be carried out again to get $T = C + A \cdot B$.

Of course, the results of the bottom-up and the top-down methods are the same. The advantage of the bottom-up method is that we also get the expressions of all events besides the top event in terms of minimal cut sets.

6.9 Dual Fault Trees

The *dual fault tree* is the logical complement of the original fault tree and is located in the success space. It is created by negating all events in the tree and substituting "union" by "intersection" and vice versa.

Negating the top event yields the minimal cut sets in the success space. For the fault tree from Fig. 6.3, the calculation would be

$$
\begin{aligned}
\overline{T} &= \overline{C + A \cdot B} \\
&= \overline{C} \cdot \overline{A \cdot B} \\
&= \overline{C} \cdot (\overline{A} + \overline{B}) \\
&= \overline{C} \cdot \overline{A} + \overline{C} \cdot \overline{B}.
\end{aligned}
\tag{6.7}
$$

So, the top event in the tree in Fig. 6.3 does not occur if either A and C or B and C do not occur.

6.10 Probability of the Top Event

Suppose we have three minimal cut sets M_1, M_2, and M_3 and want to compute the probability of the top event $P(T) = P(M_1 \cup M_2 \cup M_3)$. According to Table 6.2, this is

$$
\begin{aligned}
P(T) &= P(M_1 \cup M_2 \cup M_3) \\
&= P(M_1) + P(M_2) + P(M_3) - P(M_1 \cap M_2) - P(M_2 \cap M_3) \\
&\quad - P(M_1 \cap M_3) + P(M_1 \cap M_2 \cap M_3).
\end{aligned}
\tag{6.8}
$$

If the minimal cut sets are independent, one can obtain the following:

$$
\begin{aligned}
P(T) &= P(M_1) + P(M_2) + P(M_3) - P(M_1)P(M_2) - P(M_2)P(M_3) \\
&\quad - P(M_1)P(M_3) + P(M_1)P(M_2)P(M_3).
\end{aligned}
\tag{6.9}
$$

If the minimal cut sets are not independent, one has to look at their inner structure. For example, if A, B, and C are basic events (and therefore independent) and $M_1 = A \cap B$ and $M_2 = B \cap C$, one gets

$$
\begin{aligned}
P(M_1 \cap M_2) &= P(A \cap B \cap C) = P(A)P(B)P(C) \\
&\neq P(M_1)P(M_2) = P(A)P(B)^2 P(C).
\end{aligned}
\tag{6.10}
$$

Here one sees that it is important to assume that the basic events are independent. Otherwise, it would not be possible to compute the probability of the top event because one would have to compose the basic events which contradicts the definition of basic event.

Since looking at events is more complex than just multiplying the probabilities, we can use (6.9) to estimate the minimal computing time of $P(T)$. The estimated number of computing operations $ECT(n)$ for the probability of a top event with n minimal cut sets is

$$ECT(n) := \#\text{additions/subtractions} + \#\text{multiplications}$$

$$= \sum_{k=1}^{n-1} \binom{n}{k} + \sum_{k=1}^{n} (k-1) \binom{n}{k} = \sum_{k=1}^{n} k \binom{n}{k} - 1 = n2^{n-1} - 1, \quad (6.11)$$

where the last equality follows by differentiating the binomial theorem $\sum_{k=0}^{n} \binom{n}{k} x^k = (1+x)^n$ with respect to x and by then setting $x = 1$.

Table 6.4 shows that $ECT(n)$ increases very fast. Hence, even with modern computers it is still necessary to find good approximations for the probability of top events with many minimal cut sets. The easiest approximation is

$$P(M_1 \cup M_2 \cup \ldots \cup M_n) \approx P(M_1) + P(M_2) + \ldots + P(M_n). \quad (6.12)$$

The two sides are equal for minimal cut sets that cannot happen at the same time and the approximation is good if the probabilities of the minimal cut sets are small. This is normally true for fault trees. Because

$$P(A \cup B) = P(A) + P(B) - P(A \cap B) \leq P(A) + P(B), \quad (6.13)$$

the approximation overestimates the true probability of the top event.

Table 6.4 Number of necessary computing operations for the probability of a top event with n minimal cuts

n	$ECT(n)$
1	0
2	3
3	11
4	31
5	79
10	5119
50	$2.81 \cdot 10^{16}$
100	$6.34 \cdot 10^{31}$
500	$8.18 \cdot 10^{152}$
1,000	$5.36 \cdot 10^{303}$

Remark The number of theoretically possible minimal cut sets $n(b)$ increases exponentially with the number b of basic events:

$$n(b) := \sum_{k=1}^{b} \binom{b}{k} = 2^b - 1. \tag{6.14}$$

6.11 Importance Measures

Importance measures assign a number to each event. Comparing those numbers shows the components in which improvement is most effective. The exact meaning of the numbers depends on the importance measure. A few examples are given in the following subsections. In most cases, it turns out that only a few events influence the top event significantly. Often less than 20% of the events contribute more than 80% of the probability that the top event occurs.

6.11.1 Importance of a Minimal Cut Set

Let $M_1, M_2, \ldots, M_n$ be the minimal cut sets and T the top event. Then

$$M(M_i) := \frac{P(M_i)}{P(T)} \tag{6.15}$$

computes the relative contribution of the minimal cut set M_i to the probability of the top event.

6.11.2 Top Contribution Importance or Fussell–Vesely Importance

Next, we regard an importance measure for basic events. This importance measure is called *Fussell–Vesely importance* (also: FV importance) or "top contribution importance" (Zio et al. 2006) and is calculated by

$$FV(E_i) = \frac{P\left(\bigcup_{j \in K_i} M_j\right)}{P(T)}, \tag{6.16}$$

where K_i contains the indices of the minimal cut sets M_j that contain the event E_i (Zio 2006; Pham 2011). In particular, $FV(E_j) = M(E_j)$ for all minimal cut sets E_j of order 1.

If the minimal cut sets M_j are disjoint for $j \in K_i$, (6.16) can be written as

$$FV(E_i) := \sum_{j \in K_i} \frac{P(M_j)}{P(T)}. \tag{6.17}$$

Remark If the probabilities of the minimal cut sets are small, (6.17) is a good approximation of (6.16).

6.11.3   Risk Reduction Worth (RRW)

The "*Risk Reduction Worth* (RRW) [...] is the decrease in risk if the [component] is assumed to be perfectly reliable" (PSAM 2002). It is computed by

$$RRW(E_i) := \frac{P(T)}{P_{P(E_i)=0}(T)}, \tag{6.18}$$

where $P_{P(E_i)=0}(T)$ is the probability that the top event occurs in the modified tree where $P(E_i) = 0$ (Pham 2011).

6.11.4   Risk Achievement Worth (RAW)

The "*Risk Achievement Worth* (RAW) [...] is the increase in risk if the [component] is assumed to be failed at all times" (PSAM 2002). It is defined as

$$RAW(E_i) := \frac{P_{P(E_i)=1}(T)}{P(T)}, \tag{6.19}$$

where $P_{P(E_i)=1}(T)$ is the probability that the top event occurs in the modified tree where $P(E_i) = 1$ (Pham 2011).

6.11.5   Birnbaum Importance Measure

The *Birnbaum importance measure* computes the change in the probability of occurrence of the top event in dependence of a change in the probability of occurrence of a given event. It is usually calculated by

$$BB(E_i) := P_{P(E_i)=1}(T) - P_{P(E_i)=0}(T), \tag{6.20}$$

See, for example, (Pham 2011).

6.11.6 Measure for Risk Reduction per Investment

Measure that can be used for cost-driven component or subsystem improvement

$$BB_{\text{per cost}}(E_i) := \frac{P_{P(E_i)=p\_\text{initial}}(T) - P_{P(E_i)=p\_\text{improved}}(T)}{\Delta \, \text{cost}_i}, \tag{6.21}$$

where the right-hand side denotes the difference of the top event probability for the initial and the improved probability of failure of the basic event under consideration over the cost increase for this improvement.

6.12 Extensions of Classical Fault Tree Analysis

Classical fault tree analysis as it has been described so far cannot be used to analyze time- or mode-dependent components of complex systems. It also has the strong restriction of independent basic events and one needs sharp probabilities for the calculations. These limitations are broadly discussed in the recent literature on fault trees. The present section gives a short introduction to the ideas to extend fault tree analysis.

6.12.1 Time- and Mode-Dependent Fault Trees

A classical fault tree refers to one top event. It does not consider different modes of the system or different time intervals. The standard software does not cover these problems.

The most obvious approach to analyzing a system with different modes with standard software is to divide the occurrence of the top event into the occurrence of the top event in the different phases using an "exclusive or" gate, that is, $T = T_1 \cup_{ex} T_2$.

This approach does not yield correct results. Consider, for example, a top event of the form $T = A \cap B$. The approach yields

$$T = (A_1 \cap B_1) \cup (A_2 \cap B_2), \tag{6.22}$$

where the indices indicate in which phase an event occurs. This decomposition is correct if A and B must happen simultaneously for the top event to occur. In general, a system could also fail if A and B happen in different phases. The correct decomposition is

$$T = (A_1 \cap B_1) \cup (A_1 \cap B_2) \cup (A_2 \cap B_1) \cup (A_2 \cap B_2). \qquad (6.23)$$

Hence, in general, the approach underestimates the probability that T occurs.

On the other hand, an "inclusive or" is often used instead of an "exclusive or." This overestimates the probability that T occurs.

Remark There is no reason to assume that the underestimation and overestimation cancel each other out.

Another problem of this approach is that it increases the number of basic events which drastically increases the number of minimal cut sets, see (6.14). This, on the other hand, increases the number of computing operations exponentially, see (6.11).

Despite the problems, this approach is still used when dividing the fault tree in mode- or time-interval-dependent segments to calculate it with the standard software if specialized software is not available (Ericson 2005).

6.12.2 Dynamic Fault Tree Analysis

There is no consistent definition of *dynamic fault tree analysis* (DFTA) in the literature. It often only says that it is more complex than FTA as it is documented in the "Fault Tree Handbook" (Vesely et al. 1981).

Here DFTA is understood as FTA that uses Markov analysis to solve sequence-dependent problems, that is, problems where it is essential in which order events occur.

Example Consider a system consisting of a component and its backup and the top event "the system fails." The system only uses the backup if the component fails. If the mechanism that switches between the component and its backup fails after the component has failed and the backup has started, the top event does not occur. If, on the other hand, the switching mechanism fails before the component fails, the top event occurs. Hence, the order in which the events occur decides whether the top event occurs.

Sequence-dependent problems are normally modeled by so-called "priority and" gates which are computed by transforming Markov chains. For complex systems, this leads to a strong increase in computation time. In the literature, for example, in (Dugan and Xu 2004; Dugan et al. 2007; Huang and Chang 2007; Yuge and Yanagi 2008; Rao et al. 2009), approaches for a numerical approximation are presented.

Computing a standard fault tree by substituting the "priority and" gate by a usual "and" gate overestimates the probability of failure.

6.12.3 Dependent Basic Events

For classical FTA, independence of the basic events is essential. There have been many attempts to extend FTA to dependent basic events, see, for example, (Vesely 1970; Gulati and Dugan 1997; Andrews and Ridley 1998; Andrews and Dugan 1999; Rideley and Andrews 1999).

Most approaches keep the classical fault tree structure and introduce new gates. These gates identify the dependent sections of the tree and regard the sort of dependence. The most often applied dependency model is the Beta factor model, see (Amjad et al. 2014) for an example.

The analysis of such a modified fault tree uses the so-called Markov state transition diagrams (Ericson 2005). Here the probabilities of failure for those dependent components are being calculated and are included into the fault tree. This process is re-done until the fault tree has a structure where all basic events are independent. Now this tree can be analyzed with the standard methods.

6.12.4 Fuzzy Probabilities

Classical FTA only considers sharp probabilities. If there are inaccuracies, one assumes that inaccuracies cancel each other out throughout the analysis process but it is not totally clear that this assumption is true.

Song, Zhang, and Chan (Hua Song et al. 2009) developed a method, the so-called TS-FTA (after the Takagi–Sugeno (Takagi and Sugeno 1985) model it is based on), where events in a classical fault tree can be expressed by *fuzzy probabilities*. This works by exchanging normal gates with so-called "T-S fuzzy" gates. They do not assume the convergence of fuzzy probabilities to a sharp final result but include the convergence in the model. Furthermore, their TS-FTA model can regard the significance of an analyzing mistake for the model.

6.13 Questions

(1) For which purpose can FTA be used during development and safety validation?
(2) What are the steps to conduct an FTA?
(3) Which key concepts and construction rules should be applied during construction of an FTA?
(4) Which mathematical approaches are used to quantify an FTA?
(5) Regard the tree in Fig. 6.5.

    (a) How many minimal cut sets does this tree have?
    (b) What is the highest order of a minimal cut set in this tree?
    (c) Determine the minimal cut sets of the dual tree.

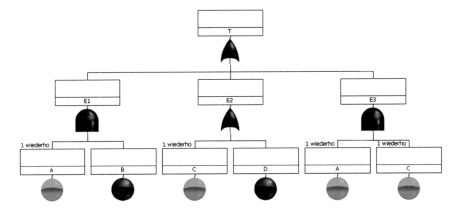

Fig. 6.5 Example fault tree generated with a software. Here "wiederho" means "repeated" and denotes basic events which occur more than once in the fault tree. They have a different color than events which occur only once. The tree was created with Relex 2011

(6) Consider the event $E = A + B + C$ where A is the event that component a fails, B the event that component b fails, and C the event that component c fails. Let $P(A) = 10^{-3}$, $P(B) = 10^{-5}$, and $P(C) = 10^{-4}$.

    (a) What is the probability of E if A, B, and C are independent?

    (b) What is the probability of E if component b fails only if component c fails and A and C are still independent?

    (c) What is a conservative estimate for the probability of E if one does not know whether A, B, and C are independent?

(7) Regard independent A, B, and C, $P(A) = 2 \cdot 10^{-3}$, $P(B) = 3 \cdot 10^{-3}$, $P(C) = 10^{-5}$, and $T = A \cdot B + C$.

    (a) Compute

        (i) the importance of both minimal cut sets, that is, $M(A \cdot B)$ and $M(C)$,

        (ii) the top contribution importance of A, that is, $FV(A)$,

        (iii) the risk reduction worth of A and C, that is, $RRW(A)$ and $RRW(C)$, and

        (iv) the risk achievement worth of A and C, that is, $RAW(A)$ and $RAW(C)$.

    (b) Interpret the results of (a).

(8) Consider the following situation (see Fig. 6.6): A train approaches a gated railroad crossing. Before reaching the crossing, it contacts a sensor. The sensor sends a signal to an engine which closes the gate.

It is assumed that any car that drives onto the tracks before the gate is completely closed drives away fast enough not to be hit. Only cars that drive onto the tracks when the gate should be closed but is not due to a failure get hit.

The following (arbitrary) frequencies are also given, see Table 6.5.

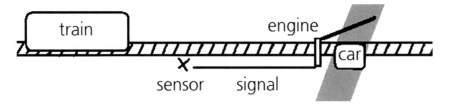

Fig. 6.6 Situation at the railroad crossing

Table 6.5 Assumed frequencies for the exercise

Situation	Frequency
Car comes at the wrong moment (independent of being stopped by a gate or not)	10^{-2}
Sensor fails	10^{-5}
Signal fails	10^{-6}
Engine fails	10^{-4}

(a) Set up a fault tree for the top event "train hits car" where all "or" gates represent disjoint unions.

(b) Apply the top-down and bottom-up methods to the fault tree from (a) to calculate the minimal cut sets.

(c) Compute the probability of the top event with the results from (b).

(d) Set up a fault tree for the top event "train hits car" with only one "or" gate.

(e) Compare the trees from (a) and (d). What are the advantages and disadvantages?

(f) Compute the minimal cut sets of the tree from (d).

(g) Compute the probability of the top event with the results from (f).

(h) How would the safety mechanism "Emergency stop of the train if there is a car on the tracks when the train arrives at the sensor" change the probability of the top event? Which assumption in the initial situation would have to be changed to make this safety mechanism very valuable?

(i) Compute the relative contribution of each minimal cut set in (d) to the probability of the top event.

(j) Compute the Fussell–Vesely importance, the RRW, the RAW, and the BB for the event $B =$"Sensor fails" using the tree from (d).

(9) The formulas for the estimated number of computing operations $ECT(n)$ have only been briefly explained in the chapter. Here they will be treated in more detail.

(a) Show that $ECT(n) = n2^{n-1} - 1$ holds for $n = 4$ in the case of independent minimal cut sets.

(b) Use (a) to explain why the second equality (6.11) looks the way it does.

Hint: $\binom{n}{k}$ = #possibilities to choose k out of n.

(c) What does $\frac{ECT(n)}{ECT(n-1)}$ converge to? What does that mean?

6.14 Answers

(1) Aims of FTA application include to compare different system designs (e.g., parallel versus more sequential designs), to determine combinations of failures that lead to top event of interest (qualitative analysis), to determine the failure probabilities of systems and subsystems (quantitative analysis), to determine critical components regarding the occurrence of the top event, to determine the most efficient improvement options of systems, to comply with standards, and to identify root causes of system failures.

(2) See Sect. 6.4.

(3) See Sects. 6.5 and 6.6.

(4) The following steps are conducted:

(i)   Graphical fault tree using well-defined symbols (gates and events);
(ii)   Use top-down or bottom-up transformations in Boolean expression;
(iii) Use Boolean algebra to simplify (e.g., absorption, distributive law, etc.);
(iv)   Use inclusion/exclusion formula for computation of top event probability; and
(v)   Use statistical independence for final quantification step.

(5) (a) 3, (b) 2,
(c) We express the tree in an equation:

$$T = E_1 + E_2 + E_3$$
$$= A \cdot B + C + D + A \cdot C$$
$$= A \cdot B + C + D,$$

where the last line follows because $A \cdot C \subseteq C$.
The equation shows that the minimal cuts are C, D, and $A \cdot B$.

(6) (a) For independent A, B, and C, we can use the formula for the probability of unions in the following way:

$$P(E) = P(A + B + C)$$
$$= P(A) + P(B) + P(C) - P(A \cdot B) - P(A \cdot C) - P(B \cdot C)$$
$$+ P(A \cdot B \cdot C)$$
$$= P(A) + P(B) + P(C) - P(A)P(B) - P(A)P(C) - P(B)P(C)$$
$$+ P(A)P(B)P(C)$$
$$= 10^{-3} + 10^{-5} + 10^{-4} - 10^{-8} - 10^{-7} - 10^{-9} + 10^{-12}$$

$$\approx 1.10989 \cdot 10^{-3}.$$

(b) If component b fails only if component c fails, then $B \subseteq C$. Hence,

$$\begin{aligned}
P(E) &= P(A + B + C)\\
&= P(A + C)\\
&= P(A) + P(C) - P(A)P(C)\\
&\approx 1.0999 \cdot 10^{-3}.
\end{aligned}$$

(c) Since $A \cdot B \cdot C \subseteq A \cdot B$ and all probabilities are bigger than or equal to 0,

$$\begin{aligned}
P(E) =& P(A + B + C)\\
=& P(A) + P(B) + P(C) - P(A \cdot B) - P(A \cdot C) - P(B \cdot C)\\
& + P(A \cdot B \cdot C)\\
\leq& P(A) + P(B) + P(C)\\
=& 1.11 \cdot 10^{-3}.
\end{aligned}$$

One can see that this estimate is quite good when comparing it to (a).

(7) (a) The importance of the minimal cut sets is computed by

(i) $M(A \cdot B) \approx \dfrac{2 \cdot 10^{-3} \cdot 3 \cdot 10^{-3}}{1.59999 \cdot 10^{-5}} \approx 0.375,$

$$M(C) \approx \dfrac{10^{-5}}{1.59999 \cdot 10^{-5}} \approx 0.625.$$

(ii) The top contribution importance of A can be derived from (a) because $FV(A) = M(A \cdot B)$.

(iii) $RRW(A) = \dfrac{P(T)}{P(C)} = \dfrac{1.59999 \cdot 10^{-5}}{10^{-5}} = 1.59999,$

$$RRW(C) = \dfrac{P(T)}{P(A \cdot B)} = \dfrac{1.59999 \cdot 10^{-5}}{6 \cdot 10^{-6}} = 2.66665.$$

(iv) $RAW(A) = \dfrac{P(A)P(B) + 1 - P(A)P(B)}{1.59999 \cdot 10^{-5}} = \dfrac{1}{1.59999 \cdot 10^{-5}}$

$$= 62500.39$$

(b) One can see that improving C is more effective than improving A.

(8) (a) The fault tree yields the following equations:

$$\begin{aligned}
T &= E_1 \cdot A,\\
E_1 &= E_2 + B,
\end{aligned}$$

$$E_2 = \overline{B} \cdot E_3,$$
$$E_3 = E_4 + C,$$
$$E_4 = \overline{C} \cdot D,$$

where $T =$ train hits car, $E_1 =$ a part fails, $A =$ car comes at the wrong moment, $E_2 =$ another part than sensor fails, $B =$ sensor fails, $E_3 =$ signal or engine fail, $E_4 =$ signal works and engine fails, $C =$ signal fails, and $D =$ engine fails.

(b) Bottom-up:

$$E_3 = \overline{C} \cdot D + C,$$
$$E_2 = \overline{B} \cdot (\overline{C} \cdot D + C) = \overline{B} \cdot \overline{C} \cdot D + \overline{B} \cdot C,$$
$$E_1 = \overline{B} \cdot \overline{C} \cdot D + \overline{B} \cdot C + B,$$
$$T = (\overline{B} \cdot \overline{C} \cdot D + \overline{B} \cdot C + B) \cdot A = \overline{B} \cdot \overline{C} \cdot D \cdot A + \overline{B} \cdot C \cdot A + B \cdot A.$$

Top-down:

$$T = E_1 \cdot A$$
$$= (E_2 + B) \cdot A = E_2 \cdot A + B \cdot A$$
$$= \overline{B} \cdot E_3 \cdot A + B \cdot A$$
$$= \overline{B} \cdot (E_4 + C) \cdot A + B \cdot A = \overline{B} \cdot E_4 \cdot A + \overline{B} \cdot C \cdot A + B \cdot A$$
$$= \overline{B} \cdot \overline{C} \cdot D \cdot A + \overline{B} \cdot C \cdot A + B \cdot A.$$

So, the minimal cut sets are $\overline{B} \cap \overline{C} \cap D \cap A, \overline{B} \cap C \cap A$, and $B \cap A$.

(c) From (b), we know that $T = \overline{B} \cdot \overline{C} \cdot D \cdot A + \overline{B} \cdot C \cdot A + B \cdot A$. All unions are disjoint, so

$$P(T) = P(\overline{B} \cdot \overline{C} \cdot D \cdot A) + P(\overline{B} \cdot C \cdot A) + P(B \cdot A).$$

All events are independent of the other events in the intersections. Hence,

$$P(T) = P(\overline{B})P(\overline{C})P(D)P(A) + P(\overline{B})P(C)P(A) + P(B)P(A)$$
$$= (1 - 10^{-5})(1 - 10^{-6})10^{-4}10^{-2} + (1 - 10^{-5})10^{-6}10^{-2} + 10^{-5}10^{-2}$$
$$\approx 1.11 \cdot 10^{-6}.$$

(d) The tree has the equations

$$T = E_1 \cdot A,$$
$$E_1 = B + C + D,$$

where

T = train hits car, E_1 = a part fails, A = car comes at the wrong moment, B = sensor fails, C = signal fails, and D = engine fails.

(e) The advantage of the tree in (a) is that all unions are disjoint, so the computation of the probabilities is easier, because $P(X + Y) = P(X) + P(Y)$ can be used. The disadvantage is that it is a more complex tree structure. (d) has a very simple and intuitive tree structure. The computation of the probability of the top event is more complex though, because

$$P(X + Y + Z) = P(X) + P(Y) + P(Z) - P(X \cdot Y) - P(X \cdot Z)$$
$$- P(Y \cdot Z) + P(X \cdot Y \cdot Z)$$

has to be used.

(f) $T = E_1 \cdot A$
$$= (B + C + D) \cdot A$$
$$= B \cdot A + C \cdot A + D \cdot A.$$

Hence, the minimal cut sets are $A \cap B$, $A \cap C$, and $A \cap D$.

(g) See Fig. 6.7 where the sprinkled area means that it belongs to the light gray area and the black area and for the big sprinkled area, it means that it belongs to the dark gray area.

As one can clearly see in Fig. 6.7, the minimal cut sets are not disjoint. So, the computation looks as follows:

$$P(T) = P(B \cdot A + C \cdot A + D \cdot A)$$
$$= P(B \cdot A) + P(C \cdot A) + P(D \cdot A)$$

Fig. 6.7 Minimal cut sets for (e)

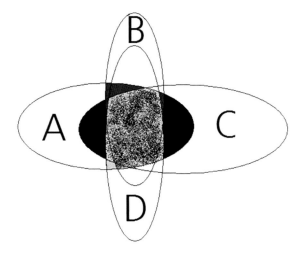

$$-P((B \cdot A) \cdot (C \cdot A)) - P((B \cdot A) \cdot (D \cdot A)) - P((C \cdot A) \cdot (D \cdot A))$$
$$+P((B \cdot A) \cdot (C \cdot A) \cdot (D \cdot A))$$
$$=P(B \cdot A) + P(C \cdot A) + P(D \cdot A)$$
$$-P(B \cdot C \cdot A) - P(B \cdot D \cdot A) - P(C \cdot D \cdot A)$$
$$+P(B \cdot C \cdot D \cdot A)$$
$$=P(B)P(A) + P(C)P(A) + P(D)P(A)$$
$$-P(B)P(C)P(A) - P(B)P(D)P(A) - P(C)P(D)P(A)$$
$$+P(B)P(C)P(D)P(A)$$
$$=10^{-5}10^{-2} + 10^{-6}10^{-2} + 10^{-4}10^{-2}$$
$$-10^{-5}10^{-6}10^{-2} - 10^{-5}10^{-4}10^{-2} - 10^{-6}10^{-4}10^{-2}$$
$$+10^{-5}10^{-6}10^{-4}10^{-2}$$
$$\approx 1.11 \cdot 10^{-6}$$

Of course, this is the same probability as in (c).

(h) The safety mechanism "Emergency stop of the train if there is a car on the tracks when the train arrives at the sensor" would not change the probability of the top event at all, because it was assumed that a car which is on the tracks before the gate is completely closed moves away quickly enough. If the cars would be slower, this safety mechanism would be interesting.

(i) $M(A \cap B) = \dfrac{P(A \cap B)}{P(T)} = \dfrac{10^{-2}10^{-5}}{1.10999 \cdot 10^{-6}} = 0.090090991,$

$M(A \cap C) = \dfrac{P(A \cap C)}{P(T)} = \dfrac{10^{-2}10^{-6}}{1.10999 \cdot 10^{-6}} \approx 0.0090090991,$

$M(A \cap D) = \dfrac{P(A \cap D)}{P(T)} = \dfrac{10^{-2}10^{-4}}{1.10999 \cdot 10^{-6}} \approx 0.90090991.$

(j) $FV(B) = \dfrac{P(A \cap B)}{P(T)} = M(A \cap B) = 0.090090991.$

By setting $P(B) = 0$ in the formula in (i), one gets

$$P_{P(B)=0}(T) = P(C)P(A) + P(D)P(A) - P(C)P(D)P(A)$$
$$= 10^{-6}10^{-2} + 10^{-4}10^{-2} - 10^{-6}10^{-4}10^{-2} \approx 1 - 01 \cdot 10^{-6}.$$

For $P(B) = 1$, one gets

$$P_{P(B)=1}(T) = P(A) + P(C)P(A) + P(D)P(A) - P(C)P(A)$$
$$- P(D)P(A) - P(C)P(D)P(A) + P(C)P(D)P(A)$$
$$= P(A) = 10^{-2}.$$

Hence,

$$RRW(B) = \frac{P(T)}{P_{P(B)=0}(T)} \approx \frac{1.11 \cdot 10^{-6}}{1.01 \cdot 10^{-6}} \approx 1.099,$$

$$RAW(B) = \frac{P_{P(B)=1}(T)}{P(T)} \approx \frac{10^{-2}}{1.11 \cdot 10^{-6}} \approx 9,009.01,$$

$$BB(B) = P_{P(B)=1}(T) - P_{P(B)=0}(T) \approx 10^{-2} - 1.01 \cdot 10^{-6} \approx 0.00999899.$$

(9) The computation and reasoning go as follows:

(a) $P(T) = P(A \cup B \cup C \cup D)$

$\quad = P(A) + P(B) + P(C) + P(D) - P(A \cap B) - P(A \cap C)$

$\quad - P(A \cap D) - P(B \cap C) - P(B \cap D) - P(C \cap D)$

$\quad + P(A \cap B \cap C) + P(A \cap B \cap D) + P(A \cap C \cap D) + P(B \cap C \cap D)$

$\quad - P(A \cap B \cap C \cap D)$

$\quad = P(A) + P(B) + P(C) + P(D) - P(A)P(B) - P(A)P(C)$

$\quad - P(A)P(D) - P(B)P(C) - P(B)P(D) - P(C)P(D)$

$\quad + P(A)P(B)P(C) + P(A)P(B)P(D) + P(A)P(C)P(D)$

$\quad + P(B)P(C)P(D) - P(A)P(B)P(C)P(D)$

Counting computing operations in the last term yields 14 additions/subtractions and 17 multiplications. Since $n2^{n-1} - 1 = 4 \cdot 2^3 - 1 = 31 = 17 + 14$, this completes the computation.

(b) By Section 6.10 we only have to regard situations where the minimal cut sets are independent because if they are not, the calculation is more difficult and the computation time is higher.

Looking at the solution of (a), there are n probabilities with one event in them. Adding them yields $n - 1$ additions, and multiplications are not necessary. Next, all possible intersections of two events are subtracted. That is, there are $\binom{n}{2}$ subtractions and $\binom{n}{2}$ multiplications. Next, all possible intersections of three events are added. That is, there are $\binom{n}{3}$ additions and $2 \cdot \binom{n}{3}$ multiplications. Lastly there is one probability with four events. This yields one subtraction and $3 = 3\binom{n}{n}$ multiplications.

In total, we have

$$n - 1 + 0 + \binom{n}{2} + \binom{n}{2} + \binom{n}{3} + 2\binom{n}{3} + 1 + 3\binom{n}{4}$$

$$= n + \binom{n}{2} + \binom{n}{3} + 0 + \binom{n}{2} + 2\binom{n}{3} + 3\binom{n}{4}$$

$$= \sum_{k=1}^{n-1} \binom{n}{k} + \sum_{k=1}^{n} (k-1)\binom{n}{k}$$

computing operations.

(c) $\dfrac{ECT(n)}{ECT(n-1)} = \dfrac{n2^{n-1} - 1}{(n-1)2^{n-2} - 1} = \dfrac{2n - 1/2^{n-2}}{(n-1) - 1/2^{n-2}}$

$$= \dfrac{2 - 1/(n2^{n-2})}{1 - 1/n - 1/(n2^{n-2})} \longrightarrow n \to \infty \dfrac{2}{1} = 2.$$

This means that if there already is a big number of minimal cut sets, every additional minimal cut set doubles the computation time.

References

Amjad, Qazi Muhammad Nouman; Zubair, Muhammad; Heo, Gyunyoung (2014). Modeling of common cause failures (CCFs) by using beta factor parametric model. In: ICESP (Hg.): International Conference on Energy Systems and Policies, Proceedings. ICESP. Islamabad, Pakistan, 24.-26.11.2014, pp. 1–6.

Andrews, J. ,D. and L. M. Ridley (1998). Analysis of systems with standby dependencies. 16th International System Safety Conference, Seattle.

Andrews, J. D. and J. B. Dugan (1999). Dependency Modeling Using Fault Tree Analysis. 17th International System Safety Conference.

Blischke, W. R. and D. N. Prabhakar Murthy (2003). Case Studies in Reliability and Maintenance, Wiley Series in Probability and Statistics.

Echterling, N., A. Thein and I. Häring (2009). Fehlerbaumanalyse (FTA) - Zwischenbericht.

Ericson, C. (2005). Hazard Analysis Techniques for System Safety, Wiley.

DIN IEC 61025 (2007). Fehlzustandsbaumanalyse (IEC 61025:2006); Fault Tree Analysis (FTA) (IEC 61025:2006). Berlin, DIN Deutsches Institut für Normung e.V., DKE Deutsche Komission Elektrotechnik Informationstechnik im DIN und VDE.

Dugan, J. B. and H. Xu (2004). Method and system for dynamic probability risk assessment. U. o. V. P. Foundation. US 20110137703 A1.

Dugan, J. B., G. J. Pai and H. Xu (2007). "Combining Software Quality Analysis with Dynamic Event/Fault Trees for High Assurance Systems Engineering." Fraunhofer ISE.

Gulati, R. and J. B. Dugan (1997). A modular approach for analyzing static and dynamic fault trees. Reliability and Maintainability Symposium. 1997 Proceedings, Annual.

Huang, C.-Y. and Y.-R. Chang (2007). "An improved decomposition scheme for assessing the reliability of embedded systems by using dynamic fault trees." Reliability Engineering and System Safety 92: 1403-1412.

Hua Song, H., H.-Y. Zhang and C. W. Chan (2009). Fuzzy fault tree analysis based on T-S model with application to INS/GPS navigation system.

Klenke, A. (2007). Probability Theory: A Comprehensive Course, Springer.

Kumamoto, H. (2007). Satisfying Safety Goals by Probabilistic Risk Assessment, Springer.

Pham, H. (2011). Safety and Risk Modeling and Its Applications, Springer.

PSAM (2002). PSA Glossary. International Conference on Probabilistic Safety Assessment and Management. San Juan, Puerto Rico USA.

Rao, K. D., V. Gopika, V. V. S. Sanyasi Rao, H. S. Kushwaha, A. K. Verma and A. Srividya (2009). "Dynamic fault tree analysis using Monte Carlo simulation in probabilistic safety assessment." Reliability Engineering and System Safety **94**: 872–883.

Relex (2007). Relex reliability Studio Getting Started Guide.

Relex (2011). Relex reliability Studio Getting Started Guide.

Rideley, L. M. and J. D. Andrews (1999). "Optimal design of systems with standby dependencies." Quality and Reliability Engineering International **15**: 103–110.

Takagi, T. and M. Sugeno (1985). "Fuzzy identification of systems and its application to modeling and control." IEEE Transactions on Systems, Man, and Cybernetics **15**(1): 116–132.

Vesely, W. E. (1970). "A time-dependent methodology for fault tree evaluation." Nuclear Engineering and Design **13**(2): 337-360.

Vesely, W. E., F. F. Goldberg, N. H. Roberts and D. F. Haasl (1981). Fault Tree Handbook. Washington, D.C., Systems and Reliability Research, Office of Nuclear Regulatroy Research, U.S. Nuclear Regulatory Comission.

Vesely, W., M. Stamatelatos, J. Duagan, J. Fragola, J. Minarick and J. Railsback (2002). Fault Tree Handbook with Aeorspace Applications. Washington, DC 20546, NASA Office of Safety and Mission Assurance, NASA Headquaters.

Yuge, T. and S. Yanagi (2008). "Quantitative analysis of a fault tree with priority AND gates." Reliability Engineering and System Safety **93**: 1577–1583.

Zio, E. (2009). Computational methods for reliability and risk analysis, World Scientific.

Chapter 7
Failure Modes and Effects Analysis

7.1 Overview

Chapter 1 introduces the inductive system analysis method *failure modes and effects analysis (FMEA)*. It presents the classical definitions of the FMEA method as well as the classical application domains on system concepts, system designs, and the production process. Such distinctions are by now often only considered as application cases, e.g., in the current German automotive association (VDA) standards.

The steps to conduct successful FMEAs are provided and discussed. It is emphasized that modern FMEAs take up many elements of similar tabular analysis such as hazard analysis as well as classical extensions such as asking for root causes, immediate consequences of failures, or a criticality assessment of failures on system level (FMECA). Some templates even ask for potential additional failures to the failure considered which would lead to catastrophic events pointing toward double failure matrix like approaches.

Modern extensions such as FMEDA are introduced, which assesses the diagnostic coverage (DC) and ultimately the safe failure fraction (SFF) of failures as required for functional safety. In all cases, an emphasis is also on valid evaluation options of obtained results sufficient for decision-making regarding system improvements. For instance, the following distinctions are made: semi-quantitative assessment of risks, of risks considering detection options and quantification of risks, respectively, of single failure events only.

When comparing with the more formal method fault tree analysis (FTA), major limitations of FMEA are discussed, in particular, that it considers single failures only, that it considers immediate effects on system level only, and is a rather formalistic cumbersome inductive approach.

The chapter will introduce the FMEA method as follows. Sections 7.2 and 7.3 focus on the method, in general, and the different steps of the execution of a classical FMEA. The form sheet that has to be filled out is described in Sect. 7.4 by discussing each row and column headline. Here, it is explained what has to be written in which column of the sheet also by providing examples.

© The Author(s), under exclusive license to Springer Nature Singapore Pte Ltd. 2021 101
I. Häring, *Technical Safety, Reliability and Resilience*,
https://doi.org/10.1007/978-981-33-4272-9_7

The chapter ends with a short summary of methods to extend classical FMEA, see Sects. 7.9 to 7.11. Examples include how to consider expert ratings and uncertainties and non-classical evaluation schemes that offer further insights into system criticality.

These last sections also take a more critical look at the method by regarding its disadvantages. Disadvantages not yet mentioned include the need of knowledgeable and iterative conduction and often lack of acceptance in development teams, in particular, for early phases of development where subsystems or function FMEAs are most advantageous.

FMEA is a good basis for safety analysis, in particular, fault tree analysis. The method can be used to analyze complex systems on the level of single elements or system functions but the team also has the possibility to define the depth of the analysis as needed, e.g., on system concept level to scrutinize ideas. An experienced and motivated team with a realistic vision or good knowledge of the system is essential for a successful FMEA.

For more details on FMEA, see, for example, the standards (VDA 2006) and (DIN 2006). The chapter is a summary, translation, and partial paraphrasing of (Hofmann 2009) in Sects. 7.2 to 7.9.3 and of (Ringwald and Häring 2010) in Sects. 7.9.4 and 7.9.5, both with additional new examples and quotations.

7.2 Introduction to FMEA

FMEA is a standard method for risk analysis that accompanies the design and development of a system. It is used to systematically detect potential errors in the design phases of the development process and prevent them with appropriate countermeasures.

Section 7.2.1 defines FMEA and presents some basic properties of the method. It is followed by a list of situations where FMEA should be applied, see Sect. 7.2.2. One distinguishes between system, construction, and process FMEA. Those three types are explained in Sect. 7.2.3.

After reading this section, one has a brief overview of the method and can look at the different steps and properties of the method in more detail.

7.2.1 General Aspects of the FMEA Method

The "International Electrotechnical Vocabulary (IEV)" defines FMEA as "a qualitative method of reliability analysis which involves the study of the fault modes which can exist in every sub-item of the item and the determination of the effects of each fault mode on other sub-items of the item and on the required functions of the item" (IEC 2008).

FMEA is an inductive system analysis method. It is used to detect, document, and prevent possible failures and their reasons during the development of a product.

The method was developed by the NASA in the 1960s for their Apollo program (Ben-Daya and Raouf 1996).

The *Rule of 10* according to which the costs to fix failures multiply by 10 in each step of the life cycle (Wittig 1993; Müller and Tietjen 2003) indicates that the costs to apply FMEA are saved in later stages of the development process of a product because with FMEA failures are detected earlier in the process.

FMEA consists of three phases which are important for the success of the analysis (Müller and Tietjen 2003), which are as follows:

(1) Determination of all system elements, possible failures, and their consequences in a group process, for example, by using a "mind map" (Eidesen and Aven 2007).
(2) Research on reliable data on frequency, severity, and detection of the failure.
(3) Modification of the system and the process design and the development of a control process that is based on the FMEA report.

7.2.2 FMEA Application Options

Nowadays, FMEA is especially spread in the automotive and electronics industry (Ringwald and Häring 2010). It is mostly used if one of the following criteria is applied (Müller and Tietjen 2003):

– New developments,
– New machinery/processes,
– Safety-relevant parts,
– Problematic parts/processes,
– Change of application area or surroundings, and
– Maintenance and repair (updates of FMEA).

7.2.3 Sorts of FMEA

Different phases of the product's development process, such as planning, development, construction, and production, ask for a different emphasis in the FMEA. Usually one distinguishes between three main types of FMEA (system FMEA, construction FMEA, and process FMEA) that build upon and complement one another (Ringwald and Häring 2010), see Table 7.1.

The formal procedure is the same for all three types and the same form sheets can be used. The form sheets are adapted to a specific type by a different interpretation of the first column of the sheet, see Sect. 7.3.8.

The three types (system FMEA, construction FMEA, and process FMEA) are explained in more detail in Sects. 7.2.3.1 to 7.2.3.3.

Table 7.1 Comparison of the three main FMEA types (Hofmann 2009)

	System FMEA Product concept FMEA	Construction FMEA Design FMEA	Process FMEA Production FMEA
Content	Superior product/system	Element/component	Production process
Requirements	Concept of the product	Design documents	Production plan
Field of responsibility	Development	Construction	General planning areas

7.2.3.1 System FMEA

System FMEA analyzes the functional interactions of the components in a system and their interfaces (Dittmann 2007). It is used in early stages of the development process to detect possible failures and weaknesses on the basis of the requirements specification and to guarantee a smooth interaction of the system's components (Müller and Tietjen 2003). By analyzing wiring and functional diagrams, the early system FMEA can guarantee the system's reliability and safety as well as the adherence of requirements.

Remark "A software *requirements specification* (SRS) is a comprehensive description of the intended purpose and environment for software under development. The SRS fully describes what the software will do and how it will be expected to perform." (TeachTarget 2012).

7.2.3.2 Construction FMEA

The *construction FMEA* starts when the design documents have been completed. The parts lists and drawings are used to analyze whether the process fulfills the construction rules of the requirements specification. This analysis takes place on the component level and examines the system in terms of developing mistakes and process mistakes in which probability of occurrence can be reduced by a change in the construction (Müller and Tietjen 2003).

The procedure of the construction FMEA is similar to the system FMEA but now specific elements or components are analyzed instead of bigger parts of the system. To avoid a too long and expensive analysis, the system is divided into functional groups and only those with new parts or problematic elements are analyzed.

Remark Sometimes system and construction FMEAs are combined to a design FMEA.

7.2.3.3 Process FMEA

The *process FMEA* finds and eliminates failures in the production process. It takes place when the production plan is completed, hence after the system and construction FMEA.

The goal of the process FMEA is the quality assurance of the final product. It has to satisfy the client's expectations and the process should be optimized and safe (Deutsche Gesellschaft für Qualität e.V. 2004; Hofmann 2009; Ringwald and Häring 2010).

The process FMEA starts with the development of a so-called process flow diagram that gives an overview of the production processes. Afterward, the flow diagram is used to examine the different work steps for possible effects of a defect. External processes such as storage or shipping can be included in this process.

7.3 Execution of an FMEA

To assure a systematic procedure, FMEA can be divided into the following five steps:

(1) Structural analysis,
(2) Functional analysis,
(3) Failure analysis,
(4) Measure analysis, and
(5) Optimization.

These steps will be explained in Sects. 7.3.2 to 7.3.6. Section 7.3.1 describes the preparations that are necessary before starting the FMEA.

7.3.1 Preparation

A team of at most eight members should be chosen to execute the FMEA. Here, it is important to have a good selection of team members. The group should include technicians or engineers from the different working areas of development, production, experiments, quality management, and sales but also clients of the product. The success depends highly on the experiences, the imagination, and the motivation of the team members although it often cannot be directly measured (Mariani 2007). Thinking and imagination are encouraged when working in such a diverse group and the necessary, extensive knowledge is assured to be present (DIN 2006; Mariani, Boschi et al. 2007).

7.3.2 Step 1: Structural Analysis

In the first step of the FMEA, the system is divided into its system elements and documented in a hierarchical structure, for example, a tree, to have a systematic overview of all components, elements, and processes (Ringwald and Häring 2010). For this, the interfaces of the system elements are determined and systems with many interactions become comprehensible (Ringwald and Häring 2010). Each element should only appear once in the diagram (Dittmann 2007). The diagram should be available to the group members during the analysis or even be developed in the group (Naude, Lockett et al. 1997).

The structure is used to determine the depth of the analysis. The depth is not fixed beforehand. In this way, it can be adapted to the system and to the assessment of the group members on how to subdivide the system (Deutsche Gesellschaft für Qualität e.V. 2004; Dittmann 2007).

The *structural analysis* hence lays the foundation for the further procedure because with the developed diagrams every element can now be regarded concerning its function or malfunction (Hofmann 2009).

7.3.3 Step 2: Functional Analysis

In the second step, every system element is analyzed regarding functionality and malfunction. For this, it is also important to have information about the environment, for example, about the temperature, dust, water, or electrical disturbances (VDA 2006).

The description of the function must be unique and comprehensive. If functions of different system elements work together, this is demonstrated in a functional structure, for example, a tree, where subfunctions of a function are logically connected within the structure. The structure also shows the interfaces of system elements. An example of such a structure is given in Fig. 7.1.

7.3.4 Step 3: Failure Analysis

The next step is *failure analysis* which is based on the system and functional structure. Here three things are analyzed (Eberhardt 2008; Hofmann 2009) as follows:

(1) Failure/malfunction: Determination of all possible failures of the regarded system element. These can be derived from the function of the system element. In particular, more than one failure can occur in a system element.
(2) Failure result: The results of a failure are the failures in superior or related system elements.

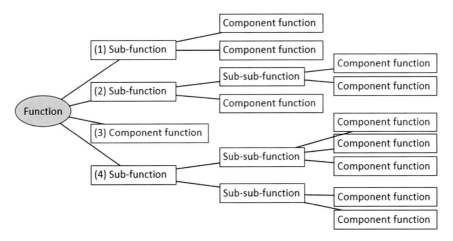

Fig. 7.1 Functional tree of a function (Hofmann 2009)

(3) Failure cause: The causes of failure can be malfunctions in subordinated or related system elements.

In this context, it is also referred to the failure types introduced in Sects. 4.10 and 4.11.

7.3.5 Step 4: Measure Analysis (Semi-quantification)

Measure analysis judges the risk and consequences of measures which are completed at the time of the analysis. Measures that have not been completed are evaluated according to the expected result and a responsible person is appointed and a deadline is set (Pentii and Atte 2002). Measures can be qualitative, semi-quantitative, or quantitative.

One distinguishes between preventive and detection measures:

Preventive measures are already achieved during the development process of a system by the right system setup. The probability of occurrence of a failure is kept small. Preventive measures should be comprehensively documented and the documentation should be referred to on the form sheet.

Detection measures find possible failures or test implemented preventive measures. Again, a comprehensive documentation is important. Detection measures should also be tested for effectiveness.

For every measure, a responsible person and a deadline should be named.

An example of documentation in this step is shown in Fig. 7.2.

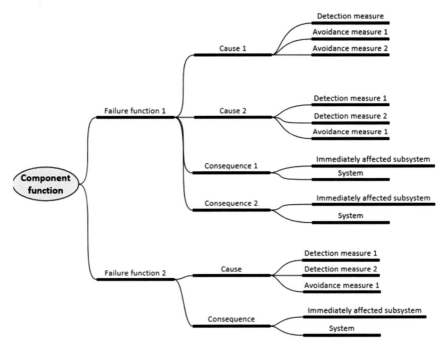

Fig. 7.2 Exemplary mind map of the malfunctions of an element, their causes, consequences on system and subsystem levels, and the established measures (Hofmann 2009)

7.3.6 Step 5: Optimization

If the results of a temporal or final evaluation are not satisfying, new measures are being proposed which are treated according to Step 4.

Step 5 is repeated until an effectiveness control shows the desired result, e.g., regarding system functions in the concept phase, regarding construction solutions in the design phase, and production process details in case of production processes.

7.4 FMEA form Sheet

7.4.1 Introduction

There is no clearly defined standard for an FMEA *form sheet*. Applicable norms suggest form sheets but those should only be understood as guidelines. The user can adapt them to his or her project's requirements. Examples for standards that include form sheets are (VDA 2006) and (DIN 2006). A further example from the North American automotive domain is (AIAG 2008).

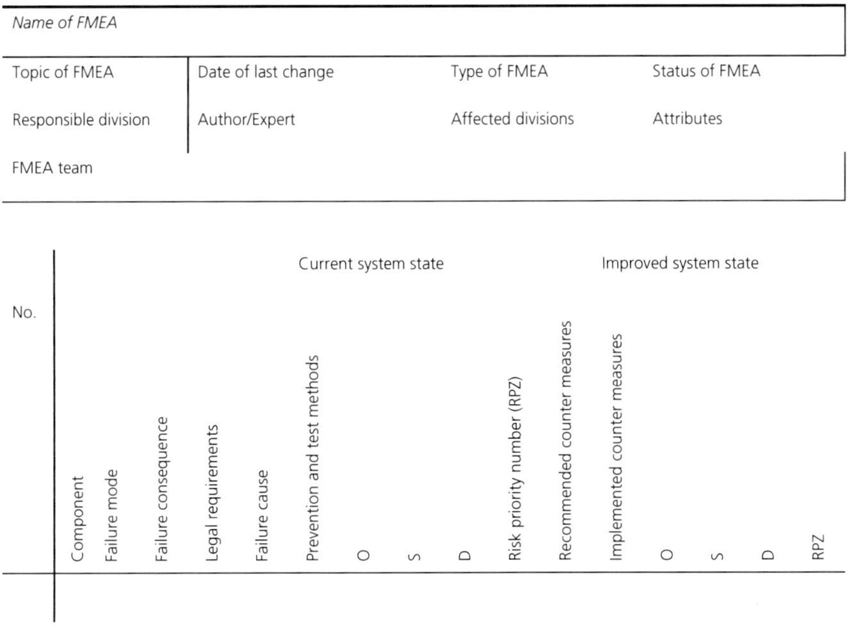

Name of FMEA			
Topic of FMEA	Date of last change	Type of FMEA	Status of FMEA
Responsible division	Author/Expert	Affected divisions	Attributes
FMEA team			

Fig. 7.3 Form sheet as introduced by (Müller and Tietjen 2003)

Figure 7.3 shows an example from the literature.

Figure 7.4 is an example from (Ringwald and Häring 2010) based on (VDA 2006). The two exemplary form sheets have a slightly different structure but contain the same information. The only major difference in terms of content is the first column ("component") in.

Figure 7.3 which is missing in

Figure 7.4, we will follow the enumeration in

Figure 7.3, because with the additional first column, it is easier to see how the table can be used for all three sorts of FMEA.

7.4.2 Columns

Here the columns are explained as numbered in Fig. 7.3, the columns in other form sheets might vary from this list but essentially contain the same information.

7.4.2.1 Column 1: Component

The first column varies, depending on the type of FMEA that is executed, see Sect. 7.2.3.

Comp any	**Failure Modes and Effects Analysis** product process	Date:	Page of
Process step:	Part/drawing (description, number, as of): Responsibility for planning and manufacturing:	Team:	Evaluation of the stated measures: Revision: Date:

Possible failure effect	S	Possible failure	Possible failure cause	Prevention/counter measures	O	Recovering measures	D	RPN	Recommended / implemented measures	Responsible / target date	Completion date	S	O	D	RPN

Fig. 7.4 Form sheet after (VDA 2006) by Ringwald and Häring (2010)

On a system FMEA form sheet, the first column contains the functions of the superior system.

On a construction FMEA form sheet, the first column contains the components of the regarded product. Information about the components can be generated from the functional block diagram, customer requirements, functional descriptions, parts lists, drawings, and wiring diagrams.

On a process FMEA form sheet, the first column contains the process steps. Information can be generated from flowcharts; the requirements specification, norms, and standards; and work sheets of comparable products.

7.4.2.2 Column 2: Failure

In Column 2, all possible failures for a component or process are named. To find possible failures, it is helpful to negate the component's function and find the plausible causes.

Example Common failures for electromechanical systems are, for example, "broken," "decline in performance," "interrupted," "short-circuit," "fatigue," "sound," "high resistance," and "corrosion."

7.4.2.3 Column 3: Failure Results

Column 3 lists all potential consequences of the failures from Column 2. Consequences should be regarded for the system, the product, and the production process and also include the client's perspective.

Remark It should be kept in mind that a failure might have more than one consequence.

7.4.2.4 Column 4: D

If the failure affects the adherence to legal requirements, this should be marked in this column. The corresponding requirement should be mentioned, for example, in a footnote.

7.4.2.5 Column 5: Failure Causes

Column 5 lists the failure causes. Again, note that a failure might have multiple causes.

7.4.2.6 Column 6: Preventive and Detection Measures

Column 6 lists preventive and detection measures, see Sect. 7.3.5. Since those measures decrease the occurrence of the failure or its causes, they are considered in the failure evaluation.

7.4.2.7 Columns 7 to 10: Evaluation

In Columns 7 to 9, the results of the failure evaluation are noted. This includes the occurrence rating O, the severity rating S, and the detection rating D. A number from 1 to 10 is assigned to each parameter. For details, see Sect. 7.5. The RPN (risk priority number) in Column 10 is the product of O, S, and D, see Sect. 7.6.

7.4.2.8 Column 11: Recommended Actions

For the failures with the biggest RPN in the computation from Sect. 7.4.2.7, recommended actions are written in Column 11. The name of the responsible person and a timeline for implementation should also be noted.

7.5   Evaluation Table

The evaluation of failures is composed of the determination of three ratings O, S, and D:

O Occurrence rating of the failure,
S Severity rating of the failure in relation to the whole system, and
D (non) detection rating of the failure.

O, S, and D take numbers between 1 and 10 where 10 stands for a probability close to 1 or high significance.

O, S, and D are determined subjectively by the team that is doing the FMEA. Hence, professional competence and experience of the team members are very important for a good evaluation. If quantitative information, for example, the occurrence probability of the failure, is known, this can also be converted into a rating number (Table 7.2).

The *occurrence rating* O can only be decreased by an improvement of the construction or the process. If $O > 8$, a more intensive investigation should follow, independent of the total RPN (Ringwald and Häring 2010).

The *severity rating* S depends on the failure results and on the view of the client who hence should be included in the evaluation. A high value of S shows a high potential damage. A severity rating $S > 8$ is called a critical characteristic. If it impairs the security or violates legal requirements, it should be analyzed in more detail, independently of the total RPN (Ringwald and Häring 2010).

The *detection rating* D judges whether the failure is noticed before the client receives the product. A detection rate $D = 9$ means that a constructional defect is only noticed during the production preparations and $D = 10$ means that the client notices the construction defect. D can be improved by changing the construction, the process, or the evaluation method (Mohr 2002).

Table 7.2 Qualitative classification of the numbers for the occurrence rating, following (Steinbring 2008)

Rating	Meaning
1	Unlikely
2	Unlikely
3	Very low
4	Very low
5	Low
6	Low
7	Moderate
8	Moderate
9	High
10	High

7.6 RPN

The risk resulting from a failure can be described by

$$R := O \cdot S. \tag{1.1}$$

Remark As in the classical definition of risk, risk is proportional to the probability of occurrence and the consequences but instead of using a probability between 0 and 1, we use the occurrence rating $O \in \{1, 2, \ldots, 10\}$. Hence, the values R taken here are higher than the values one would have computed with the classical definition of risk.

The *risk priority number* (RPN) is defined by

$$RPN := O \cdot S \cdot D. \tag{7.2}$$

By definition the risk priority number can take values between 1 and 1,000 where 1 stands for very low and 1,000 for very high risk. The RPN makes it possible to notice possible risks, rate them numerically, and distinguish big from small risks (Sony Semiconductor 2000).

A fixed bound of the RPN for a project is, in general, not considered reasonable (Müller and Tietjen 2003) because the same RPN does not mean that the risk is equal, too. We have also seen in Sect. 7.5 that investigations should follow if one of the values O, S, and D is higher than eight (Müller and Tietjen 2003; Syska 2006). So, the RPN should only be seen as an indicator that helps to prioritize the improvement measures.

Instead of using RPN itself as an indicator, it is also possible to use the Pareto method to determine critical failures. For this, the RPNs are sorted from highest to lowest. Then for every RPN one computes the fraction (Satsrisakul 2018)

$$RPN_i^{normalized} = \frac{RPN_i}{\sum_{i=1}^{n} RPN_i}, \tag{7.3}$$

that is, the RPN divided by the sum of all RPNs. The RPNs for which the fraction in (7.3) are high, are considered as critical failures, and countermeasures are decided in the team. The advantage is that the normalized RPNs for different systems in the same domain typically look similar, which allows to better motivate critical thresholds for countermeasures. Here, preventive measures for the failure cause are given more priority than detection measures (NASA Lewis Research Center 2003).

7.7   Probability of Default

In the case of a quantitative analysis, one can compute the *probability of default* of a component B by

$$PD_{B,i} = \tilde{O}_i \cdot \tilde{D}_i, \qquad (7.4)$$

where $\tilde{O}$ is the probability of occurrence of a failure, $\tilde{D}$ is the probability of detection, and the index i is the number that is assigned to the component in the FMEA table.

Remark The probability of occurrence and the probability of detection are not the same as the occurrence rating and the detection rating, see also Table 7.3. They typically are of the form $a \cdot 10^{-b}$ with $a > 0$ and $b > 1$. In Sect. 7.5, we assigned the ratings which are numbers between 1 and 10. Multiplying those ratings O and D instead of $\tilde{O}$ and $\tilde{D}$ would yield a number between 1 and 100 and instead of a probability.

If the events that a component fails are disjoint, the probability that an arbitrary component fails can be computed by

$$PD_{total} = \sum_{i=1}^{N} \tilde{O}_i \cdot \tilde{D}_i, \qquad (7.5)$$

where N is the total number of components and PD_{total} can be used as an estimate for the probability that the system fails.

Remark The computation in (7.5) is only meaningful if all components of the system are regarded in the FMEA.

Table 7.3 Probabilities of occurrence and detection following (VDA 1996) where 5E–01 is short for $5 \cdot 10^{-1}$ (Hofmann 2009)

Rating	Probability of occurrence		Detection probability
	Construction or system FMEA	Process FMEA	Construction, process, or system FMEA
10	5E–01	1E–01	1E–01
9	1E–01	5E–02	1E–01
8	5E–02	2E–02	2E–02
7	1E–02	1E–02	2E–02
6	1E–02	5E–03	2E–02
5	1E–03	2E–03	3E–03
4	5E–04	1E–03	3E–03
3	2E–04	5E–04	3E–03
2	1E–04	5E–05	3E–03
1	0	0	1E–04

The method to estimate the probability of a system failure by PD_{total} is also called Parts Count. It is the simplest and most pessimistic estimate because it is assumed that every failure of a component causes a failure of the whole system. Redundancies and the fact that some failures only cause a system failure if they occur in a special combination are not regarded.

The *total reliability* is defined as

$$Rb_{total} = 1 - PD_{total}. \qquad (7.6)$$

Remark The total reliability is often denoted by R. We use Rb to avoid confusions with R for risk.

In (Krasich 2007), Krasich notes that even if all components are analyzed in terms of their failures, the FMEA cannot make a credible statement about the reliability of the product because all failures are regarded independently of each other. The failures' interactions have to be analyzed with different methods such as Markov analysis, event tree analysis, or fault tree analysis.

7.8 Norms and Standards

There are several norms which define the approach and the implementation of FMEA. They are divided into

- Worldwide norms (ISO),
- European norms (EN),
- National norms (DIN, ÖNorm, BS, …),
- Sectoral norms (VDA 6 (automotive), GMP (pharmaceutical), …), and
- Company-specific norms (Q101 (Ford), Formel Q (Volkswagen), …).

ISO is short for "International Organization for Standardization." It was founded in 1946 and is a voluntary organization based in Geneva. Its decisions are not internationally binding treaties and the goal is rather to establish international standards. National institutions for standardization include (Hofmann 2009) the following:

- ANSI—American National Standards Institute: www.ansi.org.
- DIN—Deutsches Institut für Normung: www.din.de.
- BSI—British Standard Institute: www.bsi.org.uk.
- AFNOR—Association Française de Normalisation: www.afnor.fr.
- UNI—Ente Nazionale Italiano di Unification: www.uni.com.
- SNV—Schweizerische Normenvereinigung: www.snv.ch.
- ON—Österreichisches Normeninstitut: www.on-norm.at.

Table 7.4 lists examples of FMEA standards with different application areas.

Table 7.4 Examples of some FMEA standards and their application area

Abbreviation	Name	Application area	Year/country
EN IEC 60812	Procedures for failure mode and effect analysis (FMEA)	generic	2006/EU
BS 5760-5 Replaced by EN60812	Guide to failure modes, effects and criticality analysis (FMEA and FMECA)	generic	1991/GB
SAE ARP 5580	Recommended failure modes and effects analysis (FMEA) practices for non-automobile applications	generic, without automotive	2001/USA
MIL STD 1629	Procedures for performing a failure mode and effect analysis	military	1980/USA
VDA 4.2	Sicherung der Qualität vor Serieneinsatz	automotive	2006/D
SAE J1739	Potential failure mode and effects analysis in design (Design FMEA) and potential failure mode and effects analysis in manufacturing and assembly processes (process FMEA) and effects analysis for machinery (machinery FMEA)	automotive	2002/USA
Ford Design Institute	FMEA Handbook	automotive	2004/USA

7.9 Extensions of Classical FMEA

Section 7.9 presents some possibilities from the literature to eliminate weaknesses of classical FMEA.

The first method, based on (Chin, Wang et al. 2009) and explained in Sect. 7.9.1, makes it possible to let some of the team members have more influence on the evaluation of certain components. This can be very useful, since the teams consist of persons with different professional responsibilities.

The second method, based on (Bluvband, Grabov et al. 2004), introduces a fourth rating besides O, S, and D, the feasibility rating F.

Section 7.9.3 suggests the additional use of risk maps to not only depend on the risk priority number when deciding on improvement measures.

A short introduction to the two extensions FMECA and FMEDA of FMEA is given in Sects. 7.9.4 and 7.9.5. The C in FMECA stands for an additional criticality analysis and the D in FMEDA stands for diagnostic coverage. FMEDA determines safe failure fraction (SFF) additionally to classical FMEA as evaluation measure for the functional safety management from IEC 61508.

7.9.1 Weighting and Risk Factors

This section summarizes concepts from (Chin, Wang et al. 2009) as presented in (Hofmann 2009).

In diverse teams, the team members have different professional responsibilities. So, when rating a failure, it may be desirable that the rating of a team member with an expertise in that area weighs more than the rating of a team member whose job is related to a different part of the system.

For this, so-called *weighting factors* are introduced. For each failure, a number λ_i is assigned to the ith team member with $0 \leq \lambda_i \leq 1$ based on his or her competence in the working area related to the failure. Here, higher values for λ_i indicate more competence. The λ_i of all team members have to add up to 1, that is, for a team with n members, $\sum_{i=1}^{n} \lambda_i = 1$. The total rating is computed by

$$\text{rating} = \sum_{k=1}^{n} \lambda_k \cdot \text{rating}_k. \tag{7.7}$$

Example Consider a team consisting of a computer scientist, an engineer, and a client. Then the weighting factor for a software failure could be 0.6 for the computer scientist and 0.2 for each of the others. If the computer scientist rates the failure with 8, the engineer with 6, and the client with 9, the total rating is $0.6 \cdot 8 + 0.2 \cdot 6 + 0.2 \cdot 9 = 7.8$.

Another possible situation is that a team member would like to give more than one rating. This can be done by assigning a *risk factor* RF_i to each of his or her ratings where, like for the λ_i, $0 < RF_i < 1$ and $\sum_{i=1}^{m} RF_i = 1$ if he or she has m different ratings. The rating of team member k is

$$\text{rating}_k = \sum_{l=1}^{m} RF_l \cdot \text{rating}_{k,l}. \tag{7.8}$$

Example A team member thinks that the failure should be rated with 8 if the temperature is at least t_0 and with 7 if the temperate is below t_0 and he or she knows that statistically the temperate is at least t_0 in 60% of the time. Then his or her rating would be $0.6 \cdot 8 + 0.4 \cdot 7 = 7.6$.

The following example describes an evaluation with weighting factors and risk factors.

Example Consider a team consisting of three team members TM_1, TM_2, and TM_3. The weighting factors of the team members and their ratings including risk factors are shown in Table 7.5.

The total rating is

$$\text{rating} = \sum_{k=1}^{n} \lambda_k \sum_{l=1}^{m} RF_l \cdot \text{rating}_{k,l}$$

Table 7.5 Example of an evaluation by three team members (TM) with weighting factors and risk factors

TM$_1$, $\lambda_1 = 0.5$		TM$_2$,$\lambda_2 = 0.3$		TM$_3$, $\lambda_3 = 0.2$	
Rating	Risk factor RF	Rating	Risk factor RF	Rating	Risk factor RF
7	0.2	8	1	6	0.25
8	0.8			7	0.25
				8	0.25
				9	0.25

$$= 0.5 \cdot (0.2 \cdot 7 + 0.8 \cdot 8) + 0.3 \cdot (1 \cdot 8)$$
$$+ 0.2 \cdot (0.25 \cdot 6 + 0.25 \cdot 7 + 0.25 \cdot 8$$
$$+ 0.25 \cdot 9) = 7.8. \tag{7.9}$$

One can also compute the distribution of the ratings as displayed in Fig. 7.5. For this, one looks at one rating number and adds all products of weighting and risk factor that belong to this number. For example, for the rating 7 this is $\lambda_1 \cdot RF_{TM1/7} + \lambda_3 \cdot RF_{TM3/7} = 0.5 \cdot 0.2 + 0.2 \cdot 0.25 = 0.15 = 15\%$. In total, we have the distribution $\{(6,5\%), (7,15\%), (8,75\%), (9,15\%)\}$.

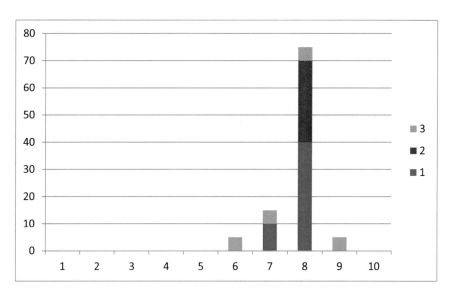

Fig. 7.5 Distribution of the ratings (Hofmann 2009)

7.9.2 Feasibility Assessment

This section summarizes concepts from (Bluvband, Grabov et al. 2004).

If there is more than one improvement measure for a system element, the RPN can be used to decide which measure should be executed.

Another interesting approach is to include a feasibility assessment in the decision. For this, the *feasibility rating F* is introduced. Similar to the classical FMEA ratings, it has a scale from 1 to 10 where 1 means that the measure is easy to implement and 10 that it is hardly feasible.

The measures are then compared by

$$\Delta_{RPN,i} = \frac{RPN_{i,\text{before measure}} - RPN_{i,\text{after measure}}}{F_i}, \qquad (7.10)$$

where the measure with the biggest value is the most effective.

7.9.3 Risk Map

As we have seen before, RPN is not the same as risk. Hence, it is sometimes useful to include both parameters into the process of deciding where improvement measures are necessary.

When including risk into the decision process, the well-known graphical presentation of risk in a risk map can be used.

The risk of each of the 100 fields in the diagram is computed by multiplying the value on the abscises with the one on the ordinate. Now each of the 100 fields in the diagram is colored green, yellow, or red, see Fig. 7.6 for an example.

Fig. 7.6 Risk map example similar as used within software FMEA applications, e.g., (Plato 2008)

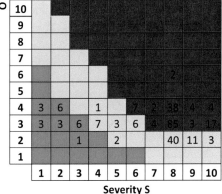

The colors have the following meaning (Helbling and Weinback 2004):

Green	Observe change in risk and probability of occurrence
Yellow	Determine responsible person, observe situation, and evaluate preventive measures
Red	Determine responsible person, develop an alternative or reaction plan, and try to reduce the probability of occurrence and the severity

The classification can be done by defining three intervals for the risk and assigning a color to each of them, for example, green for a risk of at most 9, yellow for a risk between 10 and 20, and red for a risk above 20.

It is also possible to give additional restrictions for O and S. One can see that this has been done in Fig. 7.6.

Figure 7.6 because

(a) the graph is not symmetric,
(b) $(O = 7, S = 3)$. is yellow with a risk of 21 while $(O = 10, S = 2)$ is red with a risk of 20.

The classification depends on the company and/or the project.

7.9.4 FMECA

FMECA is short for *failure modes, effects and criticality analysis*. The "International Electrotechnical Vocabulary (IEV)" defines FMEA as "a qualitative method of reliability analysis which involves a fault modes and effects analysis together with a consideration of the probability of their occurrence and of the ranking of the seriousness of the faults" (IEC 2008). That is, fault modes which are severe are not allowed to have a high probability of occurrence (Ringwald and Häring 2010).

Here, one does not rely on the subjective evaluations from Sect. 7.5 but introduces a method where absolute failure rates are determined and a key figure for the severity of a fault mode is computed. Since FMECA is based on mathematical models instead of subjective observations, the reliability of a component and the whole system can be estimated more precisely and the importance of preventive measures can be judged better (Ringwald and Häring 2010).

A so-called criticality matrix is used to display the criticality of a component. The axes are labeled with the probability of occurrence (extremely unlikely, remote, occasional, reasonably probable, frequent) and the severity (minor, marginal, critical, catastrophic) (Ringwald and Häring 2010).

An example for a criticality matrix of a component with three possible failures is given in Fig. 7.7. The colors are defined to have the following meaning:

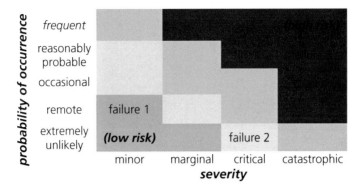

Fig. 7.7 Example for a criticality matrix. Classification from DIN EN 60812 with a sample color coding to show the evaluation of the combinations of severity class and probability class

Green	Acceptable
Yellow	*Tolerable:* Improve as low as reasonably practicable (ALARP)
Orange	*Improvement needed:* Conduct verifiable significant improvement measures
Red	Not acceptable at any rate

Note that in Fig. 7.7 events with the same risk value are assessed differently, in particular, catastrophic events. This is termed risk aversion. In a similar way, low risk priority numbers due to high detection rates only are often treated as less acceptable.

7.9.5 FMEDA

Failure modes effects and diagnostic coverage analysis (FMEDA) is another extension of FMEA. The procedure is similar to the classical FMEA but more evaluation parameters are determined (Ringwald and Häring 2010).

The safety-relevant components are analyzed in terms of criticality, diagnosis, and failure rates. The potential failures are identified and their influence on the system's safety is determined (Ringwald and Häring 2010).

According to IEC 61508, a dangerous failure is a "failure of an element and/or subsystem and/or system that plays a part in implementing the safety function that:

(a) prevents a safety function from operating when required (demand mode) or causes a safety function to fail (continuous mode) such that the EUC is put into a hazardous or potentially hazardous state; or

(b) decreases the probability that the safety function operates correctly when required" (IEC 61508 2010).

A safe failure is a "failure of an element and/or subsystem and/or system that plays a part in implementing the safety function that:

(a) results in spurious operation of the safety function to put the EUC (or part thereof) into a safe state or maintain a safe state; or
(b) increases the probability of the spurious operation of the safety function to put the EUC (or part thereof) into a safe state or maintain a safe state" (IEC 61508 2010).

Dangerous failures are further divided into those which are detected by diagnostic measures and those which stay undetected (Ringwald and Häring 2010).

The *total failure rate* λ_{total} (failures per hour) can be written as

$$\lambda_{total} = \sum \lambda_S + \sum \lambda_{DD} + \sum \lambda_{DU}, \tag{7.11}$$

where λ_S are the *safe failures* per hour, λ_{DD} are the *dangerous detected failures* per hour, and λ_{DU} the *dangerous undetected failures* per hour (Ringwald and Häring 2010).

Remark Failure rates will be introduced in more detail in Chap. 8. (Sect. 9.8) in the context of reliability prediction.

The *diagnostic coverage* (DC) is the probability that a dangerous failure is discovered in case a dangerous failure occurs through a test during system operation, e.g., production operation. It is computed by

$$DC = \frac{\sum \lambda_{DD}}{\sum \lambda_{DD} + \sum \lambda_{DU}}. \tag{7.12}$$

The total safe failure rate is the sum of all the failure rates of safe and dangerous detected failures. It is hence computed by

$$\lambda_{total/safe} = \sum \lambda_S + \sum \lambda_{DD}. \tag{7.13}$$

The *safe failure fraction (SFF)* is the ratio of the safe and detected dangerous failures to all failures. It is computed by

$$SFF = \frac{\sum \lambda_S + \sum \lambda_{DD}}{\sum \lambda_S + \sum \lambda_{DD} + \sum \lambda_{DU}} = \frac{\lambda_{total/safe}}{\lambda_{total}}. \tag{7.14}$$

SFF and DC are the evaluation parameters (as mentioned at the beginning of the section) which are added to the classical FMEA to get a FMEDA. Table 7.6 shows an example of a part of a FMEDA form sheet (Ringwald and Häring 2010).

Table 7.6 Example for the structure of a FMEDA table (Ringwald and Häring 2010)

Failure mode	Basis failure rate λ	Probability of failure	λs	λdd	λdu	DC	SFF

These failure fractions are used as a basis to determine the feasible SIL within a safety analysis following IEC 61508 by inspection of look-up best practice tables which give requirements for the possible combinations of hardware failure tolerance (HFT), SSF (or DC), and complexity of components (Level A/Level B) to reach a given intended SIL Level.

7.10 Relation to Other Methods

It is possible to use system structures, functions, and functional structures for FMEA that have been determined when executing other system analysis methods.

On the other hand, it is also possible to use, for example, a functional tree that has been developed during the FMEA as a basis for an FTA.

So, if different teams are analyzing the same system with different methods, they should compare their work to avoid doing the same analysis twice and to profit from each other.

7.11 Disadvantages of FMEA

Carrying out an FMEA takes a lot of time and has to be done by people with experience in that work field. Hence, the personnel costs are not negligible.

If an FMEA is only done because the client wishes it and not because it is a usual part of the development process, and then sometimes not by a group of persons with a lot of experience but by a single person who is available, it often does not yield useful, cost-saving results. In those cases, the FMEA report is a waste of time and money and increases neither quality, safety, or safety of the product nor does it lead to a higher client satisfaction (Teng and Ho 1996).

If the FMEA does not include flowcharts or other process diagrams, or if those are not explained to the team members, the team sometimes does not understand the system well enough to see critical functions and potential problems.

Another problem if the FMEA does not include process diagrams is that joining elements (soldered, adhesive, welded, and riveted joints, screws) are easily forgotten in the analysis (Raheja 1981).

In classical FMEA, the probability of occurrence, the severity, and the probability of detection are rated subjectively, and with that the risk priority number. Hence, there is no objective way to measure the efficiency of the improvement measures.

7.12 Questions

(1) Characterize FMEA and related similar analyses.
(2) In which steps is an FMEA conducted?
(3) What are typical columns used in an FMEA and an FMEDA?
(4) What are the main advantages and disadvantages of an FMEA?
(5) How is an FMEA semi-quantified and evaluated?
(6) Assign the three types of FMEA "system FMEA," "construction FMEA," and "process FMEA" to the columns of Table 7.7 from (Müller and Tietjen 2003).
(7) Table 7.8 shows ratings for four team members.
(a) Compute the total rating.
(b) The team decides that team member 2 is more important than they originally thought, so they want to increase his ranking by 0.1. Are they allowed to?
(8) An FMEA team consists of three members: an engineer, a sales manager, and a client. They analyze the fault that the produced color of their product does not match the planned color due to a calculation mistake in the color mixture. The planned color was originally determined by a survey.
     Which weighting factors would you use for the three team members? Why?
(9) An FMEA team consists of three members: an engineer, a sales manager, and a client. They analyze the fault that the produced color of their product does not match the planned color due to a calculation mistake in the color mixture. The planned color was originally determined by a survey.
     Should one of the persons use risk factors? In which situation?

Table 7.7 Extract of an FMEA table in (Müller and Tietjen 2003)

Type of FMEA			
Failure effect	Starter failure	Armature shaft broken	Engine does not start
Failure mode	Armature shaft broken	Too high compression of armature shaft	Starter failure
Failure cause	Too high compression of armature shaft	Production tolerances, type of jointing	Armature shaft broken

Table 7.8 Sample ratings for team members within the context of conduction of an FMEA

Team member	Rating	Weighting factor
1	9	0.3
2	5	0.1
3	6	0.2
4	6	0.4

7.13 Answers

(1) See Chap. 5. Typically, FMEA is a semi-quantitative inductive tabular approach. FMEDA is in addition quantitative.

(2) See Sect. 7.3.

(3) See Figs. 7.3 and Fig. 7.4; Tables 7.6 and 7.7.

(4) See, in particular, Sects. 7.10 and 7.11.

(5) See Sects. 7.6, 7.7, 7.9.3, and 7.9.5.

(6) Second column design level FMEA, third column process FMEA, and forth column system FMEA.

(7) (a) The total rating is $0.3 \cdot 9 + 0.1 \cdot 5 + 0.2 \cdot 6 + 0.4 \cdot 6 = 6.8$.

    (b) Only if they decrease the other ratings to have a total sum of 1.

(8) For A and E (O and D), the engineer should have the highest weighting factor, for B (S) the lowest. The engineer can best judge the production process while the sales management and the client are better in judging the impact on the consumer's behavior.

(9) They could, for example, use risk factors if they can see two different scenarios that might occur. Then the risk factor would be the estimate how likely they think the scenarios will occur.

References

AIAG (2008): Potential failure mode and effects analysis (FMEA). Reference manual. 4th ed. [Southfield, MI]: Chrysler LLC, Ford Motor Co., General Motors Corp.

Ben-Daya, M. and A. Raouf (1996). "A revised failure mode and effects analysis model." International Journal of Quality & Reliability Management, **13**(1): 43–47.

Bluvband, Z., P. Grabov and O. Nakar (2004). "Expanded FMEA (EFMEA)." Annual Reliability and Maintainability Symposium: 31–36.

Chin, K.-S., Y.-M. Wang, G. K. K. Poon and J.-B. Yang (2009). "Failure mode and effects analysis using a group-based evidential reasoning approach." Computers & Operations Research **36**(6).

Deutsche Gesellschaft für Qualität e.V. (2004). FMEA - Fehlermöglichkeits- und Einflussanalyse. Berlin, Beuth.

DIN (2006). Analysetechniken für die Funktionsfähigkeit von Systemen - Verfahren für die Fehlzustandsart- und -auswirkungsanalyse (FMEA) (IEC 60812:2006), Beuth Verlag GmbH.

Dittmann, L. U. (2007). "Fehlermöglichkeits- und Einflussanalyse (FMEA)." OntoFMEA: Ontologiebasierte Fehlermöglichkeits- und Einflussanalyse: 35–73.

Eberhardt, O. (2008). Gefährdungsanalyse mit FMEA: Die Fehler-Möglichkeits- und Einfluss-Analyse gemäß VDA-Richtlinie, Expert Verlag.

Eidesen, K. and T. Aven (2007). "An evaluation of risk assessment as a tool to improve patient safety and prioritise the resources." Risk, Reliability and Societal Safety: 171–177.

Helbling and Weinback (2004). "Human-FMEA." Diplomarbeit.

Hofmann, B. (2009). Fehlermöglichkeits- und Einflussanalyse (FMEA). EMI Bericht E 04/09, Fraunhofer Ernst-Mach-Institiut.

IEC (2008). "International Electrotechnical Vocabulary Online."

IEC 61508 (2010): Functional Safety of Electrical/Electronic/Programmable Electronic Safety-related Systems Edition 2.0. Geneva: International Electrotechnical Commission.

Krasich (2007). "Can Failure Modes and Effects Analysis Assure a Reliable Product." Reliability and Maintainability Symposium, 277-281.

Mariani, B., Colucci (2007). "Using an innovative SoC-level FMEA methodology to design in compliance with IEC61508." Proceedings of the conference on Design, automation and test in Europe, 492–497.

Mariani, R., G. Boschi and F. Colucci (2007). "Using an innovative SoC-level FMEA methodology to design in compliance with IEC61508, EDA Consortium." Proceedings of the conference on Design, Automation and Test in Europe: 492–497.

Mohr, R. R. (2002). "Failure Modes and Effects Analysis."

Müller, D. H. and T. Tietjen (2003). FMEA-Praxis, Hanser.

NASA Lewis Research Center. (2003). "Tools of Reliability Analysis - Introduction and FMEAs." Retrieved 2015-09-15, from http://www.fmeainfocentre.com/presentations/dfr9.pdf.

Naude, P., G. Lockett and K. Holmes (1997). "A case study of strategic engineering decision making using judgmental modeling and psychological profiling." IEEE TRANSACTIONS ON ENGINEERING MANAGEMENT 44(3): 237–247.

Pentii, H. and H. Atte (2002). "Failure Mode and Effects Analysis of Software- Based Automation Systems." SÄTEILYTURVAKESKUS STRÅLSÄKERHETSCENTRALEN RADIATION AND NUCLEAR SAFETY AUTHORITY **STUK-YTO-TR 190.**

Plato. (2008). "Plato - Solutions by Software." from www.plato.de.

Raheja, D. (1981). "Failure Mode and Effect Analysis - Uses and Misuses." ASQC Quality Congress Transactions - San Francisco: 374–379.

Ringwald, M. and I. Häring (2010). Zuverlässigkeitsanalyse der Elektronik einer Sensorplatine. Praxisarbeit DHBW Lörrach/ EMI-Bericht, Fraunhofer Ernst-Mach-Institiut.

Satsrisakul, Yupak (2018): Quantitative probabilistic safety assessment of autonomous car functions with Markov models. Master Thesis. Universität Freiburg, Freiburg. INATECH.

Sony Semiconductor. (2000). "Quality and Reliability HandBook." from http://www.sony.net/Products/SC-HP/quality/index.html.

Steinbring, F. (2008). "Sicherheits - und Risikoanalysen mit Verwendung der Platosoftware." Vortrag Unterlüß, 11.2008, RMW GmbH.

Syska, A. (2006). Fehlermöglichkeits- und Einflussanalyse (FMEA). Produktionsmanagement: Das A-Z wichtiger Methoden und Konzepte für die Produktion von heute, Gabler: 46–49.

TeachTarget. (2012). "SearchSoftwareQuality." Retrieved 2012-03-26, from http://searchsoftwarequality.techtarget.com/definition/software-requirements-specification.

Teng and Ho (1996). "FMEA An integrated approach for product design and process control." International Journal of Quality & Reliability Management 13(5): 8–26.

VDA (1996). "Sicherung der Qualität vor Serieneinsatz - System-FMEA."

VDA (2006). Bd.4 System FMEA im Entwurf Test 2.

Wittig, K.-J. (1993). Qualitätsmanagement in der Praxis. Stuttgart, Teubner.

Chapter 8
Hazard Analysis

8.1 Overview

In contrary to FTA, which focuses on the identification of root causes of failures on system level, and to FMEA, which focuses on effects of failures of components or basic functionalities on system level, hazard analysis focuses on hazards (potential sources of high risks, hazard sources, hazard modalities, kinds of hazards, and hazard modes) and their resulting risks on system level.

The hazard analysis can be considered as inductive when starting from the hazards and respective initial hazard lists as defined in 4.5. However, it allows for the summary of bottom-up inductive as well as top-down deductive system analysis results. In particular, it allows considering combinations of failures. This holds since it asks for the probability of hazard (modes) independent of how they occur. The relation of hazard analysis to other approaches is shown in Sect. 8.2. It shows that hazard analysis on system level is well suited to communicate system safety at management level as well as to identify the necessity of analyses, to collect the results and to ask for more refinements.

Section 8.3 describes the hazard log which is used for documentation of past failure events and their implications on system design and assessment with various means. It is updated during system analysis. Sections 8.4–8.8 explain the different types of hazard analyses. They are used at different points in time during system developmentas they need more and more but also different information:

- Preliminary hazard list (PHL, Sect. 8.4),
- Preliminary hazard analysis (PHA, Sect. 8.5),
- Subsystem hazard analysis (SSHA, Sect. 8.6),
- System hazard analysis (SHA, Sect. 8.7), and
- Operation and support hazard analysis (O&SHA, Sect. 8.8).

Section 8.9 summarizes the worksheets of different types of hazard analyses by comparing which entries are needed for the different analyses.

© The Author(s), under exclusive license to Springer Nature Singapore Pte Ltd. 2021   127
I. Häring, *Technical Safety, Reliability and Resilience*,
https://doi.org/10.1007/978-981-33-4272-9_8

Section 8.10 is dedicated to the evaluation of hazard analysis. This comprises any risk comparison approaches and risk criteria, in particular, such criteria that are system or domain specific. Often used approaches are risk acceptance matrices or risk graphs. The latter can also be used to determine the necessity and requirements for safety functions, in particular, SIL levels.

Section 8.11 relates the different types of hazard analyses to the phases of the Safety Life Cycle of IEC 61508 and to a general system development cycle.

Section 8.12 mentions a standardization approach documented by Häring and Kanat (2013) for a highly safety–critical domain. This approach summarizes best practices regarding mainly for HA, FMEA, FTA, reliability prediction with standards, and hazard log. The main contribution to the chapter is how to sort and how to link analysis methods, in particular, all the tabular approaches.

Questions and answers are given in Sects. 8.14 and 8.15, respectively.

Sections 8.3–8.11 are a summary, translation, and partial paraphrasing of parts of Echterling (2011) and is mainly based on Ericson (2005). Further, Fraunhofer EMI sources for this chapter include (Echterling 2010; Thielsch 2012; Häring and Kanat 2013).

8.2   Relation to Other Methods and Order of Conduction of Different Types of Hazard Analysis

Figure 8.1 shows the relationship between the different hazard analysis methods. The abbreviations stand for hazard analysis (HA), preliminary hazard list (PHL), preliminary hazard analysis (PHA), subsystem hazard analysis (SSHA), system hazard

Fig. 8.1 Relationship between various safety analysis methods (Häring and Kanat 2013)

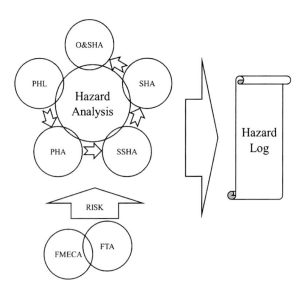

Table 8.1 Input from other analysis forms

	HL	PHL	PHA	FMEA	FTA	SSHA	SHA	O&SHA
HL		X	X	X	X	X	X	X
PHL								
PHA		X						
SSHA			X	X	X			
SHA			X	X	X	X		
O&SHA			X				X	

analysis (SHA), and operating and support hazard analysis (O&SHA). FTA and FMECA are already known from Chaps. 6 and 7.

Depending on the product, the procedure of the hazard analysis can also follow different schemes, for example, the PHA and HA can be combined for very simple systems.

Table 8.1 shows which form of safety analysis needs input from which other form. It reads as follows: the analysis method on the left side uses information from the methods in the columns marked with an "X".

8.3 Hazard Log

The *hazard log* (HL) is a continuously updated system-specific dataset of hazards. It documents the safety management for each hazard (Ministry of Defence 2007).

Hazard is understood as defined in Sect. 3.6. In the context of the hazard analysis, it can be helpful to regard the list of hazards contained in the norm DIN EN ISO 12100 (Deutsches Institut für Normung 2011). The list of hazards is not complete but it can give an overview of potential hazards (Thielsch 2012).

The hazard log itself is not a safety analysis; it is only used for documentation. It is useful in case of change of staff, of technical modernization, or to verify executed measures.

"It is a continually updated record of the hazards. It includes information documenting risk management for each hazard and accident (Ministry of Defence 2007)." (Häring and Kanat 2013).

The worksheet should at least contain the following columns (Echterling 2011):

(1) Unique identification of the hazard for references, for example, a running number;
(2) Operation mode;
(3) Hazard (title);
(4) Hazard source;
(5) Cause/trigger;
(6) Consequences;

(7)   Risk category at the beginning;
(8)   Countermeasures;
(9)   Risk category after executing the countermeasures;
(10) Verification (if necessary with links to the corresponding documents);
(11) Remark; and
(12) Date and status of work.

Remarks

– The points (8) to (12) are updated according to the measures.
– A hazard log is only added to. Entries are never deleted.
– It has to be clear who is responsible for writing the hazard log (assuring consistency).

8.4   Preliminary Hazard List (PHL)

The *preliminary hazard list* (PHL) is the starting point for the following hazard analyses. The goal is a temporary list of hazards. The estimates are conservative, that is, risks are rather over than underestimated.

The information necessary to execute a PHL and the results of a PHL are displayed in Table 8.2

The procedure of the PHL is as follows (Echterling 2011):

(1) Detection of potential hazards by brainstorming, checklists for hazard sources, and expertise.
(2) Update of the hazard log.

The head of the worksheet should at least contain the following entries (Echterling 2011):

(1) The system under analysis.
(2) Name of the person executing the analysis.
(3) Date of the execution.

The worksheet should at least contain the following columns (Echterling 2011):

(1) Unique identification of the hazard for references, for example, a running number.
(2) Operation mode.
(3) Hazard (title).
(4) Hazard source.
(5) Cause/trigger.
(6) Consequences.
(7) Comments.

Table 8.2 Tabular overview on types of hazard analyses: name, input, main analysis process tasks, and output. Based on Ericson (2005), Echterling (2010)

Type of analysis	Input	Hazard analysis tasks	Output
Preliminary hazard list (PHL)	• Design knowledge • Generic hazard lists, e.g., from standards • Experiences	• Comparison of design and hazards • Comparison of design and experiences • Documentation process	• Specific hazards • Safety–critical factors • Hazard sources
Preliminary hazard analysis (PHA)	• Design knowledge • General hazards • PHL • Top events	• Analysis of the PHL and the top events • Identification of new hazards • Evaluation of the hazards as far as the knowledge of the design allows it • Documentation process	• Specific hazards • Safety–critical factors • Hazard sources • Methods for risk reduction • System requirements
Subsystem hazard analysis (SSHA)	• Design • Specific hazards • PHA • Safety–critical functions • Gained experiences	• Evaluation of the causes of all hazards • Execution of necessary analyses • Documentation process	• Hazard sources • Risk evaluation • System requirements
System hazard analysis (SHA)	• Design • Specific hazards • PHA, SSHA • Safety–critical functions • Gained experiences	• Identification of subsystem interface hazards • Evaluation of the causes of all hazards • Execution of necessary analyses • Documentation process	• System interface hazards • Hazard sources • Risk evaluation • System requirements
Operating and support hazard analysis (O&SHA)	• Design • System operation • Handbooks • Hazard lists	• Gathering of all tasks and steps of a procedure • Evaluation of all tasks and steps • Use of identified hazards and checklists • Identification and evaluation of all hazards • Documentation process	• Specific hazards • Hazard sources • Risk evaluation • System requirements • Precautionary measures, warnings, procedures

Remarks

- The differentiation between hazard source and cause can, for example, be the following: The hazard source is explosive and the cause/trigger of the hazard is vibration.
- The analysis must be started early in the developing process.
- All potential hazards must be documented. (This includes hazards which are judged as harmless.)
- The procedure of identifying potential hazards should be well structured.
- A checklist for hazard sources must be used.
- A list of the used hardware, the functions, and the modes of operation should be available.
- System-specific abbreviations and terms should be avoided.
- The description of the hazard in the worksheet should not be too detailed but meaningful.
- There should be enough explanatory text in the worksheet.
- The description of a cause has to be clearly related to the corresponding hazard.

An extract of an exemplary German PHL is given in Table 8.3.

Table 8.3 Extract of an exemplary German PHL based on DIN EN ISO 12100 (Thielsch 2012)

Type of hazard/hazard source	Cause/Triggering event	Potential damage effects/consequences
Mechanical hazard	– Acceleration/deceleration (kinetic or rotational energy) – Moving parts – Cutting parts – Sharp edges – Rotating parts	– Squeezing – Impact – Cutting (off) – Roping in/wrapping/catching – Drawing in – Chafing/excoriation – Cropping – Puncture – Penetration or perforation
Electrical hazard	– Electric overload – Short circuit	– Injuring or fatal electro shock – Fire
Hazards caused by substances/agents	– Smoke gas/vapor – CBRN agents	– Breathing difficulties – Poisoning
Physical stress	– Stress – Mental/psychological overload – Working tasks beyond scope of expert knowledge or skills of employee	– Human error – Burnout
Human physical hazard	– Too high physical stress – Carrying of heavy loads	– Joint damage – Herniated disk

8.5   Preliminary Hazard Analysis

The *preliminary hazard analysis* (PHA) analyzes hazards, their causes, their consequences, the related risks, and countermeasures to reduce or eliminate those risks.
It is executed when detailed information is not yet available. "A mishap risk index,
see e.g. (MIL-STD-882D 2000), describes the associated severity and probability of
each hazard" (Häring and Kanat 2013). The information from the PHL is used in the
PHA.

The information necessary to execute a PHA and the results of a PHA are displayed
in Table 8.2, see third row.

The procedure of the PHA is as follows (Echterling 2011):

(1)   Copying entries from the PHL.
(2)   Identification of new potential hazards, their causes, and the related risks.
(3)   Estimation of the risk category (qualitatively or quantitatively).
(4)   Suggestion of measures to eliminate hazards or reduce risks (it is possible that
      no measures can yet be suggested at this point in time).
(5)   Evaluation of the planned measures to ensure that they meet the expectations
      and possibly suggestions for later verification procedures for the measures.
(6)   Update of the hazard log.

The PHA is a living document and transforms into a hazard analysis later (typically
after a "Preliminary Design Review").

In the following, the entries/columns that were also part of the PHL are written
in gray. Here, information from the PHL can be copied but it has to be updated. The
columns are not left out to emphasize the importance of regarding them again under
consideration of new system knowledge.

The head of the worksheet should at least contain the following entries (Echterling
2011):

(1)   The system under analysis.
(2)   Part or function of the system under analysis.
(3)   Name of the person executing the analysis.
(4)   Date of the execution.

The worksheet should at least contain the following columns (Echterling 2011):

(1)   Unique identification of the hazard for references, for example, a running
      number.
(2)   Operation mode.
(3)   Hazard (title).
(4)   Hazard source.
(5)   Cause/trigger.
(6)   Consequences.
(7)   Estimate of the risk category at the beginning (baseline).
(8)   All countermeasures.

 (9) Estimated risk category after executing the countermeasures.
(10) Planned verification procedure.
(11) Comments.
(12) Status of the hazard (the processing is either still open or completed).

Remarks

– The remarks for the PHL still hold.
– The description of the severity classes must be clearly related to the consequences.
– The probability of occurrence must be clearly related to the causes.
– The countermeasures must be clearly related to the causes and hazard sources.
– In most cases, it cannot be expected that the severity class is reduced by countermeasures.
– The risk category should be evaluated in terms of improvement after the execution of countermeasures.
– The handling of the hazard can only be finished in the framework of the PHA if the safety requirements are fulfilled.

8.6 Subsystem Hazard Analysis

The *subsystem hazard analysis* (SSHA) is executed as soon as there is enough information available about the subsystem and as soon as the system is so complex that the analysis is regarded as useful.

The goal of the SSHA is to identify causes of and countermeasures for previously determined hazards and the verification of safety on the subsystem level. Risk estimates are specified with fault tree analysis or FMEA.

For the SSHA, it is especially important to complement the tables with enough explanatory text.

"The SSHA stays unfinished as long as the risk is higher than the requirements demand" (Häring and Kanat 2013).

The information necessary to execute a SSHA and the results of a SSHA are displayed in Table 8.2.

The procedure of the SSHA is as follows (Echterling 2011):

(1) Copying entries from the PHA.
(2) Identification of new potential hazards within the subsystem.
(3) Determination of the risk category, for example, with FMEA or fault tree analysis.
(4) Determination of measures to eliminate hazards or reduce risks.
(5) Evaluation of the planned measures to ensure that they meet the expectations.
(6) Verification of the final risk category after executing all measures.
(7) Update of the hazard log.

Again, as before: In the following, the entries/columns that were also part of the PHA are written in gray. Here, information from the PHA can be copied but it has to be updated. The columns are not left out to emphasize the importance of regarding them again under consideration of new system knowledge.

The head of the worksheet should at least contain the following entries (Echterling 2011):

(1) The system under analysis.
(2) Parts or subsystem or function of the system under analysis.
(3) Name of the person executing the analysis.
(4) Date of the execution.

The worksheet should at least contain the following columns (Echterling 2011):

(1)   Unique identification of the hazard for references, for example, a running number.
(2)   Operation mode.
(3)   Hazard (title).
(4)   Hazard source.
(5)   Cause/trigger.
(6)   Consequences.
(7)   Risk category at the beginning (baseline).
(8)   All countermeasures.
(9)   Risk category after executing the countermeasures.
(10) Planned verification procedure or executed verification with reference to the corresponding document.
(11) Comments.
(12) Status of the hazard (the processing is either still open or completed).

Remarks

– The remarks for the PHL and PHA still hold.
– The SSHA must be executed within the boundaries of the subsystem.
– Detailed references are necessary, in particular, to other system analysis methods.
– The SSHA must analyze the system in as much detail as necessary for the respective hazard. It is the most detailed hazard analysis.
– The handling of hazards can only be finished when the remaining risk is acceptable and the safety requirements are fulfilled.

8.7 System Hazard Analysis

The *system hazard analysis* (SHA) evaluates risks and ensures the adherence to safety rules on system level. It is based on the PHA and possibly the SSHA. "It considers interfaces between subsystems. The SHA includes common cause failures and risks with regard to interface hazards. [...] The SHA is the hazard analysis with the widest scope. The analysis must be comprehensible by a third party. Hazards may only be marked as closed when the effectiveness of recommended actions has been verified and the safety requirements have been met" (Häring and Kanat 2013).

The information necessary to execute a SHA and the results of a SHA are displayed in Table 8.2.

The procedure of the SHA is as follows (Echterling 2011):

(1) Copying entries from the PHA.
(2) Identification of further hazards on system level.
(3) Identification of hazards from interfaces between subsystems.
(4) Determination of a risk category for the identified hazards before the execution of measures (baseline). The risk category usually is the same as in the PHA.
(5) Determination of measures to reduce risks.
(6) Monitoring of the execution of the suggested measures.
(7) Evaluation of the planned measures to ensure that they meet the expectations.
(8) Verification of the final risk category after executing all measures.
(9) Update of the hazard log.

The SHA is a living document and is typically presented at the "Critical Design Review."

Again, as before: In the following, the entries/columns that were also part of the PHA are written in gray. Here, information from the PHA can be copied but it has to be updated. The columns are not left out to emphasize the importance of regarding them again under consideration of new system knowledge.

The head of the worksheet should at least contain the following entries (Echterling 2011):

(1) The system under analysis.
(2) Name of the person executing the analysis.
(3) Date of the execution.

The worksheet should at least contain the following columns (Echterling 2011):

(1) Unique identification of the hazard for references, for example, a running number
(2) Operation mode.
(3) Hazard (title).
(4) Hazard source (e.g., interface, subsystem).
(5) Cause/trigger.
(6) Consequences.
(7) Risk category at the beginning.
(8) All countermeasures.
(9) Risk category after executing the countermeasures.
(10) Planned verification procedure or executed verification with reference to the corresponding document.

(11) Comments.
(12) Status of the hazard (the processing is either still open or completed).

Remarks

– The remarks for the PHL and PHA still hold.
– Hazards can only be marked as completed if the effectiveness of the measures is verified.
– Failures with common cause and dependent events have to be sufficiently regarded.
– Fault trees do not replace hazard analyses but do preliminary work for them.
– The handling of hazards can only be finished when the remaining risk is acceptable and the safety requirements are fulfilled.
– Detailed references are necessary.
– The analysis must be understandable for a qualified third party.

8.8 Operating and Support Hazard Analysis

The *operating and support hazard analysis* (O&SHA) captures hazards that can occur in operation and maintenance. The goal is to identify causes, consequences, risks, and measures for risk reduction.

The information necessary to execute an O&SHA and the results of an O&SHA are displayed in Table 8.2.

The procedure of the O&SHA is as follows (Echterling 2011):

(1) Making a list of all activities that are relevant in the framework of an O&SHA (this list can normally be generated from existing documents).
(2) Evaluation of all activities related to a hazard and comparison with a checklist for hazard sources.
(3) Determination of a risk category for the identified hazards before and after the execution of measures.
(4) Documentation of all measures.
(5) Checking that all necessary warnings are written in the documentation.
(6) Monitoring of the execution of the suggested measures.
(7) Update of the hazard log.

In the following, the entries/columns that were also part of the SHA are written in gray. Other than before, here information is not copied from the previous analyses. The O&SHA can, however, be combined with the SHA to a more extended SHA.

The head of the worksheet should at least contain the following entries (Echterling 2011):

(1) The system under analysis.
(2) Operation mode of the system.
(3) Name of the person executing the analysis.

(4) Date of the execution.

The worksheet should at least contain the following columns (Echterling 2011):

(1)   Unique identification of the hazard for references, for example, a running number.
(2)   Description of the procedure.
(3)   Hazard (title).
(4)   Hazard source (e.g., interface, subsystem).
(5)   Cause/trigger.
(6)   Consequences.
(7)   Risk category at the beginning.
(8)   All countermeasures.
(9)   Risk category after executing the countermeasures.
(10) Planned verification procedure or executed verification with reference to the corresponding document.
(11) Comments.
(12) Status of the hazard (the processing is either still open or completed).

Remarks

– All activities must be listed.
– All activities must be completely described.

8.9   Comparison of the Hazard Analysis Worksheets

Table 8.4 summarizes and compares the columns of the worksheets of the different types of hazard analysis.

8.10   Evaluation of Risks

The hazard analysis demands a risk evaluation for the identified hazards. The goal is to analyze whether the identified risk is below a given level of acceptance. The risk is measured by a combination of probability of occurrence and the extent of damage. The risk evaluation can be qualitative, quantitative, or semi-quantitative.

A quantitative evaluation expresses risk in terms of numbers, for example, €/year. A qualitative evaluation uses terms like "low" and "high." The semi-quantitative evaluation assigns relative numbers to each risk class, for example, numbers from 1 to 10.

Table 8.4 Entries in the work sheet of the different types of hazard analysis

	Hazard log (HL)	Preliminary hazard list (PHL)	Preliminary hazard analysis (PHA)	Subsystem hazard analysis (SSHA)	System hazard analysis (SHA)	Operating and support hazard analysis (O&SHA)
System under analysis		X	X	X	X	X
System component under analysis			X	X		
State of operation of system						X
Executing person		X	X	X	X	X
Date of the execution		X	X	X	X	X
Identification	X	X	X	X	X	X
Mode of operation	X	X	X	X	X	X
Description of procedure						X
Hazard	X	X	X	X	X	X
Hazard source	X	X	X	X	X	X
Cause	X	X	X	X	X	X
Consequences	X	X	X	X	X	X
Risk category at beginning	X		X estimate	X	X	X
Countermeasures	X					
All countermeasures			X	X	X	X
Risk category after countermeasures	X		X estimate	X	X	X
Verification procedure	X		X	X	X	X
Comments	X	X	X	X	X	X
Date and status	X		X	X	X	X

8.10.1   Risk Map

An established method for the evaluation of risks is the so-called risk map or risk matrix which has already been introduced in Sect. 7.9.3.

Figure 8.2 shows a semi-quantitative risk matrix according to MIL-STD-882D. Each combination of damage and probability of occurrence leads to one risk class.

Risk matrix		Extent of damage			
		Catastrophc	Critical	Marginal	Negligible
Probability of occurrence		Material damage over 1,000,000 €	200,000 € ≤ material damage ≤ 1,000,000 €	10,000 € ≤ material damage ≤ 200,000 €	2,000 € ≤ material damage ≤ 10,000 €
Probability level	per lifetime	I	II	III	IV
Frequent A	$P>10^{-1}$			7	13
Probable B	$10^{-1}{\geq}P>10^{-2}$			9	16
Occasional C	$10^{-2}{\geq}P>10^{-3}$		6	11	18
Remote D	$10^{-3}{\geq}P>10^{-6}$	8	10	14	19
Improbable E	$10^{-6}{\geq}P$	12	15	17	20

Fig. 8.2 Risk matrix example as proposed by MIL-STD-882D (Department of Defence 2000)

The acceptance level of each class is marked with a color. Red is high, orange serious, yellow medium, and green low risk.

The decision which field is marked in which color depends on the situation.

8.10.2 Risk Graph

Risk graphs are a semi-quantitative method to evaluate risks. The risk is estimated with a decision tree typically using three or four risk parameters (depending on which norm is used).

The IEC 61508 uses the four risk parameters to determine technical reliability requirements for safety functions. At least, the first two and the last decision options can be used for the assessment of any risk.

(C) Consequence risk parameter,

(F) Frequency and exposure time risk parameter,

(P) Possibility of failing to avoid hazard risk parameter, and.

(W) Probability of the unwanted occurrence (event frequency of triggering (initial) hazard event), resulting in the risk graph shown in Figs. 8.3 and 8.4.

"a, b, c, d, e, f, g, h represent the necessary minimum risk reduction. The link between the necessary minimum risk reduction and the safety integrity level is shown in the table" (IEC 61508 2010) (Table 8.5).

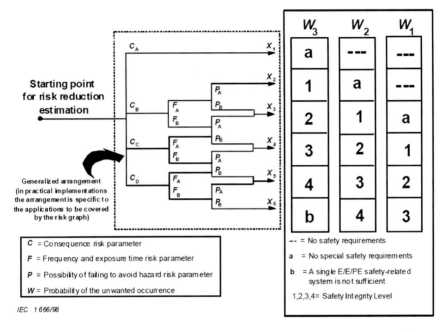

Fig. 8.3 Risk graph: general scheme (IEC 61508 2010). Copyright © 2010 IEC Geneva, Switzerland. www.iec.ch

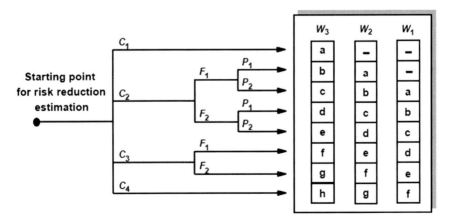

Fig. 8.4 Risk graph—example (illustrates general principles only) (IEC 61508 2010). Copyright © 2010 IEC Geneva, Switzerland. www.iec.ch

Table 8.5 Explanation of the parameters in Figs. 8.4 and 8.5 (IEC 61508 2010). Copyright © 2010 IEC Geneva, Switzerland. www.iec.ch

	Necessary minimum risk reduction	Safety integrity level
C = Consequence risk parameter		
F = Frequency and exposure time risk parameter	–	No safety requirements
P = Possibility of avoiding hazard risk parameter	a	No special safety requirements
W = Probability of unwanted occurrence	b, c	1
a, b, c … h = Estimates of the required risk reduction for the safety-related systems	d	2
	e, f	3
	g	4
	h	An E/E/PE safety-related system is not sufficient

Fig. 8.5 Standardization process (Häring and Kanat 2013)

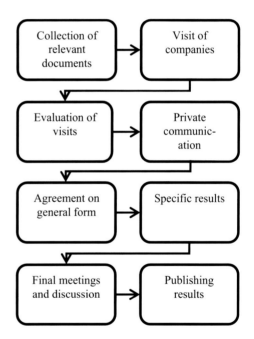

Tables 8.6 and 8.7 show examples of calibrations of the parameters C, F, P, and W for the example in Fig. 8.4 and the generic risk graph in Fig. 8.3, respectively.

The present risk graph is a visualization option for the successive consideration of influencing factors relevant for the frequency/probability and the consequences. Typically, each level means an increase or decrease by an order of magnitude (factor of 10 or 0.1). The risk graph implicitly executes an individual as well as collective risk criteria. Typically, it uses a single individual risk criterion with some deviations for large consequence numbers. In the present case, we have

$$R = WFPC \leq R_{\text{crit}}, \tag{8.1}$$

Table 8.6 Example of data relating to risk graph in Fig. 8.4 (IEC 61508 2010). Copyright © 2010 IEC Geneva, Switzerland. www.iec.ch

Risk parameter		Classification	Comments
Consequence (C)	C_1	Minor injury	1 The classification system has been developed to deal with injury and death to people. Other classification schemes would need to be developed for environmental or material damage
	C_2	Serious permanent injury to one or more persons; death to one person	2 For the interpretation of C_1, C_2, C_3, and C_4, the consequences of the accident and normal healing shall be taken into account
	C_3	Death to several people	3 See comment 1 above
	C_4	Very many people killed	4 This parameter takes into account
Frequency of, and exposure time in, the hazardous zone (F)	F_1	Rare to more often exposure in the hazardous zone	– operation of a process (supervised (i.e., operated
	F_2	Frequent to permanent exposure in the hazardous zone	– by skilled or unskilled persons) or unsupervised); – rate of development of the hazardous event (for example, suddenly, quickly, or slowly);
Possibility of avoiding the hazardous event (P)	P_1	Possible under certain conditions	– ease of recognition of danger (for example, seen immediately, detected by technical measures, or detected without technical measures);
	P_2	Almost impossible	– avoidance of hazardous event (for example, escape routes possible, not possible, or possible under certain conditions);
Probability of the unwanted occurrence (W)	W_1	A very slight probability that the unwanted occurrences will come to pass and only a few unwanted occurrences are likely	– actual safety experience (such experience may exist with an identical EUC or a similar EUC or may not exist)
	W_2	A slight probability that the unwanted occurrences will come to pass and few unwanted occurrences are likely	5 The purpose of the W factor is to estimate the frequency of the unwanted occurrence taking place without the addition of any safety-related systems (E/E/PE or other technology) but including any other risk reduction measures
	W_3	A relatively high probability that the unwanted occurrences will come to pass and frequent unwanted occurrences are likely	6 If little or no experience exists of the EUC, or the EUC control system, or of a similar EUC and EUC control system, the estimation of the W factor may be made by calculation. In such an event, a worst-case prediction shall be made

Table 8.7 Example of calibration of the general-purpose risk graph in Fig. 8.3 (IEC 61508 2010). Copyright © 2010 IEC Geneva, Switzerland. www.iec.ch

Risk parameter		Classification	Comments
Consequence (C)			1 The classification system has been developed to deal with injury and death to people Other classification schemes would need to be developed for environmental or material damage
Number of fatalities	C_A	Minor injury	
This can be calculated by determining the numbers of people present when the area exposed to the hazard is occupied and multiplying by the vulnerability to the identified hazard	C_B	Range 0,01 to 0,1	2 For the interpretation of C_A, C_B, C_C, and C_D, the consequences of the accident and normal healing shall be taken into account
	C_C	Range > 0,1 to 1,0	3 See comment 1 above
The vulnerability is determined by the nature of the hazard being protected against. The following factors can be used:	C_D	Range > 1,0	4 P_A should only be selected if all the following are true:
$V = 0,01$ Small release of flammable or toxic material			– facilities are provided to alert the operator that the SIS has failed;
$V = 0,1$ Large release of flammable or toxic material			– independent facilities are provided to shut down such that the hazard can be avoided or which enable all persons to escape to a safe area;
$V = 0,5$ As above but also a high probability of catching fire or highly toxic material			– the time between the operator being alerted and a hazardous event occurring exceeds 1 h or is definitely sufficient for the necessary actions
$V = 1$ Rupture or explosion			5 The purpose of the W factor is to estimate the frequency of the hazard taking place without the addition of the E/E/PE safety-related systems
Occupancy (F)		Rare to more often exposure in the hazardous zone. Occupancy less than 0,1	If the demand rate is very high the SIL has to be determined by another method or the risk graph recalibrated. It should be noted that risk graph methods may not be the best approach in the case of applications operating in continuous mode (see 3.5.16 of IEC 61508–4)
This is calculated by determining the proportional length of time the area exposed to the hazard is occupied during a normal working period	F_A		
	F_B	Frequent to permanent exposure in the hazardous zone	6 The value of D should be determined from corporate criteria on tolerable risk taking into consideration other risks to exposed persons
NOTE 1 If the time in the hazardous area is different depending on the shift being operated then the maximum should be selected			
NOTE 2 It is only appropriate to use F_A where it can be shown that the demand rate is random and not related to when occupancy could be higher than normal. The latter is usually the case with demands which occur at equipment start-up or during the investigation of abnormalities			

(continued)

Table 8.7 (continued)

Risk parameter		Classification	Comments
Possibility of avoiding the hazardous event (P) if the protection system fails to operate	P_A	Adopted if all conditions in column 4 are satisfied	
	P_B	Adopted if all the conditions are not satisfied	
Demand rate (W)	W_1	Demand rate less than 0,1 D per year	
The number of times per year that the hazardous event would occur in absence of a the E/E/PE safety-related system	W_2	Demand rate between 0,1 D and D per year	
To determine the demand rate it is necessary to consider all sources of failure that can lead to one hazardous event. In determining the demand rate, limited credit can be allowed for control system performance and intervention. The performance which can be claimed if the control system is not to be designed and maintained according to IEC 61508 is limited to below the performance ranges associated with SIL 1	W_3	Demand rate between D and 10 D per year For demand rates higher than 10 D per year higher integrity shall be needed	

NOTE This is an example to illustrate the application of the principles for the design of risk graphs. Risk graphs for particular applications and particular hazards will be agreed with those involved, taking into account tolerable risk, see Clauses E.1 to E.6

Table 8.8 Overview of norms and standards treating risk graphs, translated extract of Table 10.3 in Thielsch (2012)

Standard	Title	Year
ISO 13849–1	Safety of machinery–safety-related parts of control systems–Part 1: General principles for design	2008
IEC 61508	Functional safety of electrical/electronic/programmable Electronic safety-related systems	2010
IEC 61511	Functional safety–safety instrumented systems for the process industry sector	2005
VDI/VDE 2180	Safeguarding of industrial process plants by means of process control engineering	2010

where R_{crit} can be derived for each combination of the risk graph. Typically, it is the same for small consequence numbers and decreases for larger consequence numbers. Often the factor C rates fatalities 10 times more than injuries. Another option to use the same risk criterion is to rate for instance 10 fatalities 100 times more than 1 fatality, thus implementing risk aversion.

The IEC 61508 is not the only norm using risk graphs. An extract of an overview of norms and standards treating risks graphs from Thielsch (2012) is shown in Table 8.8.

The presented risk graph approach is very generic and can also be applied to safety and security-critical systems, e.g., if the safety risks originate from security hazards (cyber-induced functional safety issues). However, an even more generic approach would be to assess the resiliency of technical systems rather than only their safety and security.

This would allow to consider in a more explicit and systematic way not only the effects of the resilience (catastrophe, disruptive event) management phases preparation, prevention, and protection on risks quantification but also much more explicitly the phases response and recovery. On top level, this outlines a possible way to resilient socio-technical systems, see e.g., (Häring 2013a, c; Häring et al. 2016).

8.10.3 Computation of SIL

The IEC 61508–5 introduces a method to quantitatively determine the safety integrity levels (SIL). The following key steps of the method are necessary for every E/E/PE safety-related system (IEC 61508 S+ 2010):

- "determine the tolerable risk from a table as [Table 8.9];
- determine the EUC [equipment under control] risk;
- determine the necessary risk reduction to meet the tolerable risk;

Table 8.9 Example of a risk classification of accidents, graphically adapted, content from IEC 61508 S+ (2010). Copyright © 2010 IEC Geneva, Switzerland. www.iec.ch

	Consequence			
Frequency	Catastrophic	Critical	Marginal	Negligible
Frequent	I	I	I	II
Probable	I	I	II	III
Occasional	I	II	III	III
Remote	II	III	III	IV
Improbable	III	III	IV	IV
Incredible	IV	IV	IV	IV
Class I	Intolerable risk			
Class II	Undesirable risk, and tolerable only if risk reduction is impracticable or if the costs are grossly disproportionate to the improvement gained			
Class III	Tolerable risk if the cost of risk reduction would exceed the improvement gained			
Class IV	Negligible risk			

– allocate the necessary risk reduction to the E/E/PE safety-related systems, other technology safety-related systems and other risk reduction measures" (IEC 61508 S+ 2010).

Here, Table 8.9 should be seen as an example what such a table can look like. It should not be used in this way since precise numbers, for example, for the frequencies, are missing (IEC 61508 S+ 2010).

In this method, the limit of a safety function of an E/E/PE safety-related system for a given level of damage is computed by

$$\text{PFD}_{\text{avg}} \leq \frac{F_t C}{F_{np} C} = \frac{F_t}{F_{np}}, \tag{8.2}$$

IEC 61508 S+ (2010), Thielsch (2012) where

PFD_{avg} is "the average probability of failure on demand of the safety-related protection system, which is the target failure measure for safety-related protection systems operating in a low demand mode of operation" (IEC 61508 S+ 2010),

F_t is the frequency of occurrence of tolerable hazards (Thielsch 2012),

F_{np} is the frequency of occurrence of the associated risk class without regarding protection measures (Thielsch 2012), and.

C the consequences (amount of damage) (Thielsch 2012).

A system's F_{np} can be determined by Thielsch (2012).

– analysis of failure rates in similar situations,
– data from relevant databases, or
– computation using appropriate prediction methods.

Table 8.10 Comparison risk matrix, translated from Thielsch (2012)

Frequency of event during operation of system per year		Consequence				
		Negligible	Marginal	Severe	Critical	Catastrophic
Incredible	0,000001	I-A	I-B	I-C	I-D	I-E
Improbable	0,00001	I-A	I-B	I-C	I-D	II-E
Remote	0,0001	I-A	I-B	I-C	II-D	II-E
Occassional	0,001	I-A	I-B	II-C	III-D	III-E
Probable	0,01	I-A	II-B	III-C	III-D	III-E
Frequent	0,1	II-A	III-B	III-C	III-D	III-E

Remark If the SIL determined with (8.2) is too high, the probability of occurrence can be reduced by appropriate protection measures. Afterward, formula (8.2) can be used again with F_p instead of F_{np} where

$$F_p = F_{np} p_1, \tag{8.3}$$

and p_1 is the probability that the protection measure prevents a failure (Thielsch 2012).

Example As mentioned before, Table 8.9 cannot be used in this way because the frequencies do not have numerical values. For the following example, we use Table 8.10 from Thielsch (2012) from whom we also have this example. This table should not be copied for projects besides this example.

We consider the example of a safety function of a safety-related E/E/PE system with low demand rate. Experts determined that in case of a failure "severe" damage would occur. The frequency of such a failure was estimated with $F_{np} = 1.5 \cdot 10^{-2} a^{-1}$.

Using Table 8.10 one can see that the risk class is III-C. This is a non-tolerable risk (red). The highest tolerable frequency for severe consequences is $F_t = 10^{-4} a^{-1}$ (the frequency belonging to the last green area in the column "severe").

Hence, the highest possible PFD_{avg} is computed by

$$\text{PFD}_{\text{avg}} = \frac{F_t}{F_{np}} = \frac{10^{-4} a^{-1}}{1.5 \cdot 10^{-2} a^{-1}} \approx 6.7 \cdot 10^{-3}. \tag{8.4}$$

We remember the SIL table from Table 3.5 that we display here again as Table 8.11 to get a better overview.

We can see that for a system with low demand rate, $\text{PFD}_{\text{avg}} = 6.7 \cdot 10^{-3}$ corresponds to SIL 2.

The experts also found out that it is possible to install an independent non-functional protection measure which reduces the probability of occurrence by 50 percent. Hence,

$$F_p = 1.5 \cdot 10^{-2} \cdot 0.5 = 7.5 \cdot 10^{-3}. \tag{8.5}$$

This leads to

Table 8.11 The SILs for high and low demand rates as defined in IEC 61508 (2010). Copyright © 2010 IEC Geneva, Switzerland. www.iec.ch

Low demand rate		High demand rate	
Per request		Per hour	
SIL	Average probability of a dangerous failure	SIL	Average frequency of a dangerous failure
4	$[10^{-5},\ 10^{-4})$	4	$[10^{-9}h^{-1},\ 10^{-8}h^{-1})$
3	$[10^{-4},\ 10^{-3})$	3	$[10^{-8}h^{-1},\ 10^{-7}h^{-1})$
2	$[10^{-3},\ 10^{-2})$	2	$[10^{-7}h^{-1},\ 10^{-6}h^{-1})$
1	$[10^{-2},\ 10^{-1})$	1	$[10^{-6}h^{-1},\ 10^{-5}h^{-1})$

$$\text{PFD}_{\text{avg}} = \frac{10^{-4}}{7.5 \cdot 10^{-3}} \approx 1.3 \cdot 10^{-2} \tag{8.6}$$

which corresponds to SIL 1 according to Table 8.11.

8.10.4 Determination of SIL Based on Individual and Collective Risk Criteria

Alternatively to the SIL determination based on risk map of Sect. 8.10.1, risk graph of Sect. 8.10.2, and the computational method of Sect. 8.10.3, the necessary SILs of systems can also be computed from first principles in terms of allowed local individual risk and collective risk criteria. An example for a vehicle protection system is provided in Kaufman and Häring (2011).

More generally, any method suited for assessing overall system safety in terms of overall acceptable risk of a system considering short-term and long-term options before, during, and/or after events can be used to determine functional safety requirements. See, for instance, the often data-driven and learning envisioned approaches for critical infrastructure systems in Bruneau et al. (2020) the process and methods listed in Häring et al. (2017) for critical infrastructure systems, the approach in Häring (2015) mainly for hazardous (explosive) events, and the methods proposed in Häring (2013b, c) for civil crisis response systems.

Example processes to be conducted range from basic in Dörr and Häring (2008) to current risk analysis and management processes (Häring 2015), performance-based risk and resilience assessment (Häring et al. 2017) to panarchy process-based approaches (Häring et al. 2020).

The application of a tabletop analysis to identify challenging risks for mass gatherings is provided in Siebold et al. (2015). Such approaches are more efficient if they are supported with software, see, e.g., (Baumann et al. 2014) for an implementation example developed within the EU project (BESECURE 2015). An example for a semi-quantitative risk and resilience scorecard-based approach for critical infrastructure assesment was developed in the EU project (RESILENS 2018), which is also suited for the identification of necessary risk mitigation measures.

Example methods for simulative risk quantification of one or more static explosive sources are (Häring et al. 2019; Salhab et al. 2011), in case of humanitarian demining (Schäfer et al. 2014b; D-BOX 2016; Schäfer and Rathjen 2014a), for (general) terroristic events with unknown locations in urban areas (Voss et al. 2012; Fischer et al. 2012a, b, 2015, 2018, 2019; Riedel et al. 2014, 2015; Vogelbacher et al. 2016) and for moving explosive sources with known trajectories in air (Häring et al. 2009). In all cases, the risk quantification approaches can be used to determine functional safety requirements, e.g., for warning and far-out and close-in protective measures.

An overview on applications in the domain of building safety and security of public spaces is given in Stolz et al. (2017). Related EU projects include (VITRUV 2014; Encounter 2015; MABS 15), and (EDEN 2016).

An example for system performance-based risk and resilience assessment for telecommunication grids is given in Miller et al. (2020), Fehling-Kaschek et al. (2019, 2020), which allows to determine safety and security issues that need to be addressed, e.g., with safety functions. An example for s similar approach for local electrical supply grids on urban quarter level is conceptualized in Tomforde et al. (2019), OCTIKT (2020). A simulative dynamic simulation for gas grid supply grids at European scale is provided in Ganter et al. (2020), which is suited to model also the effects of disruptions.

With much less but more complicated nodes, airport checkpoints can be simulated to assess the achieved level of physical safety post checks (Renger et al. 2015; XP-DITE 2017) or identification security (Jain et al. 2020).

An example of perceived safety and security assessment of persons on urban spaces is provided (Vogelbacher et al. 2019), which can be used to determine potential need for (comfort) safety functions, e.g., to switch on public street lamps if a person approaches.

8.11 Allocation of the Different Types of Hazard Analysis to the Development Cycle

Table 8.12 shows how the different types of hazard analysis can be assigned to the phases of the Safety Life Cycle of IEC 61508. A clear allocation to one phase is not always possible. Table 8.13 shows how the different types of hazard analysis can be assigned to a more general description of a development cycle.

Table 8.12 Allocation of the different types of hazard analysis to the phases of the Safety Life Cycle of IEC 61508, following (Echterling 2011)

Abbreviation	Method	Safety life cycle phase of IEC 61508[a]
PHL	Preliminary hazard list	1 Concept 2 Overall scope definition
PHA	Preliminary hazard analysis	3 Hazard and risk analysis
SSHA	Subsystem hazard analysis	7 Overall safety validation planning 9 Realization 10 Safety-related systems: other technology—Realization 11 External risk reduction facilities—Realization
SHA	System hazard analysis	7 Overall safety validation planning 9 Realization 10 Safety-related systems: other technology—Realization 11 External risk reduction facilities—Realization 13 Overall safety validation
O&SHA	Operation and support hazard analysis	6 Overall operation and maintenance planning 9 Realization 10 Safety-related systems: other technology—Realization 11 External risk reduction facilities—Realization 13 Overall safety validation

[a]For more details on the Safety life cycle according to IEC 61508 see Chap. 11

Table 8.13 Allocation of the different types of hazard analysis to a general development cycle (Echterling 2011)

Method	Concept	Preliminary design		Design		Test	Operation
PHL	X		*Preliminary Design Review*		*Critical Design Review*		
PHA		X					
SSHA				X			
SHA				X		X	
O&SHA				X		X	

8.12 Standardization Process

The following standardization approach from Häring and Kanat (2013) can be seen as a motivation to think about the safety analysis processes used in your company.

In the example analyzed in the paper, Häring and Kanat found out that "very different approaches [to safety analysis] were used. Even within the same company, safety analysis techniques often vary significantly from project to project" (Häring and Kanat 2013). Since in a lot of companies, projects are worked on where the company is only sub-contractor or where they get public money, this situation might occur more often because "depending on the project set-up, the industry has sometimes to work together with third-party auditors which require different levels of rigor of analysis" (Häring and Kanat 2013). The standardization approach developed in Häring and Kanat (2013) is shown in Fig. 8.5.

"It was decided to agree on a list of safety analysis techniques and to specify in detail how they were to be executed. This entailed a short description of the technique, a formal portrait of the process, a list of columns for tabular techniques and a catalogue of important issues that need to be considered" (Häring and Kanat 2013).

8.13 Tabular Summary of Use of Different Types of Tabular Analyses

The chapter introduced different types of hazard analysis. They lay their focus on different aspects of the system and are executed when different amounts of information are available. Table 8.14 summarizes the minimal table content of the different methods, compare also Table 8.4.

We have seen (see Table 8.4) that there are a lot of identical common columns for the different types of hazard analysis. These, however, might still be filled with different information because their analyses are executed at different time points in the development cycle (see Tables 8.12 and 8.13).

8.14 Questions

(1) Discuss major pros and cons of hazard analyses.
(2) How are HL, PHL, PHA, SSHA, SHA, and O&SHA related to each other? Draw a diagram using arrows labeled with "use for" and "update."
(3) The risks in a PHL usually are …

    (a) Precise,
    (b) Overestimated, and
    (c) Underestimated.

(4) How does the PHL differ from the hazard analyses PHA, SSHA, SHA, and O&SHA?
(5) Which of the hazard analyses is the most detailed?

Table 8.14 Analysis methods and minimal required table columns (Häring and Kanat 2013)

Columns	Analysis methods						
	PHL	PHA	SSHA	SHA	O&SHA	FMECA	HL
Unique identifier	X	X	X	X	X	X	X
Element						X	
Failure mode						X	
Failure rate						X	
Immediate effects						X	
System effect						X	
Operating mode	X	X	X	X	X		X
Hazard	X	X	X	X	X		X
Source	X	X	X	X	X		X
Cause/trigger	X	X	X	X	X	X	X
Initial risk		X	X	X	X		X
Recommended action		X	X	X	X	X	X
Residual risk		X	X	X	X		X
Verification procedure		X	X	X	X		X
Result	X	X	X	X	X		X
Comment	X	X	X	X	X	X	X
Date and status		X	X	X	X		X

(6) Is a SHA still necessary if SSHA has been done for every subsystem? Why? Why not?

(7) What is the major difference in the general idea of the SHA and the O&SHA? Name an example for a hazard and a hazard source that could appear in both types of analysis. What would be a possible cause for the hazard to appear in the SHA? What would be a possible cause for the hazard to appear in the O&SHA?

(8) How can you plan the tabular scope of hazard analyses during system development? Name typical column headings of PHA and SHA.

8.15 Answers

(1) Major advantages of hazard analyses include the following: (i) hazard analyses can be used in all phases of the development process to assess safety of systems; (ii) it covers detail up to management level; (iii) systematic inductive approach on level of kinds of hazards; (iv) it also comprises deductive approach as it asks for origins of potential hazard events; (v) it can be used to summarize and coordinate the need for other more detailed system analysis methods; and (vi) hazard log allows to summarize past experiences in the safety development process and lessons learnt in the safety management process.

Fig. 8.6 Relationship of the different types of hazard analysis. The arrow with the filled peak means "use for" and the "normal" arrow means "update"

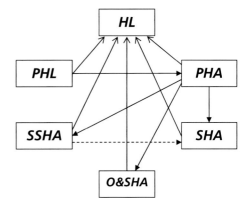

Major disadvantages include the following: (i) danger of formal conduction, similar as discussed for FMECA, see Sect. 7.11. (ii) Single probability of hazard event factor needs to consider options of prevention and mitigation before, during, and after threat events. (iii) Similar argument for damage assessment. (vi) System-specific critical novel hazard types might not be part of initial hazard lists. (v) Combination of hazard effects is not considered explicitly. (vi) Needs to resort to more advanced methods for quantification.

(2) See Fig. 8.6 where the arrows pointing to "HL" are "update" and the ones with the filled arrow head are "use for." The dashed line means "possibly."

(3) (b) Overestimated.

(4) It does not give risk categories and countermeasures, it only lists hazards.

(5) SSHA, see remarks in Sect. 8.6.

(6) Yes, it considers the interfaces between the subsystems.

(7) The SHA deals with hazards regarding the system itself. The O&SHA deals with hazards regarding the operation and maintenance of the system.

> Hazard: Gas poisoning,
> Hazard source: Leak in the gas tank,
> Cause SHA: Material defect in gas tank, and
> Cause O&SHA: Person damaging the gas tank.

(8) See Table 8.14.

References

Baumann, D; Häring, I; Siebold, U; Finger, J (2014): A web application for urban security enhancement. In Klaus Thoma, Ivo Häring, Tobias Leismann (Eds.): Proceedings/9th Future Security - Security Research Conference. Berlin, September 16 - 18, 2014. Fraunhofer-Verbund Verteidigungs- und Sicherheitsforschung; Future Security Research Conference; Future Security 2014. Stuttgart: Fraunhofer Verl., pp. 17–25.

BESECURE (2015): EU project, Best practice enhancers for security in urban environments, 2012–2015, Grant agreement ID: 285222. Available online at https://cordis.europa.eu/project/id/285222, checked on 9/27/2020.

Bruneau, Michael; Cimellaro, Gian-Paolo; Didier, Max; Marco, Domaneschi; Häring, Ivo; Lu, Xilin et al. (2020): Challenges and generic research questions for future research on resilience. In Zhishen Wu, Xilin Lu, Mohammad Noori (Eds.): Resilience of critical infrastructure systems. Emerging developments and future challenges. Boca Raton, FL: CRC Press, Taylor & Francis Group (Taylor and Francis series in resilience and sustainability in civil, mechanical, aerospace and manufacturing engineering systems), pp. 1–42.

D-BOX (2016): EU Project, Demining too-box for humanitarian clearing of large scale area from anti-personal, landmines und cluster munitions, 2013–2016, Grant agreement ID: 284996. Available online at https://cordis.europa.eu/project/id/284996, checked on 9/27/2020.

Department of Defence (2000). MIL-STD-882D Standard practice for system safety.

Deutsches Institut für Normung (2011). DIN EN ISO 12100. Sicherheit von maschinen - Allgemeine Gestaltungsgrundsätze - Risikobeurteilung und Risikominderung.

Dörr, Andreas; Häring, Ivo (2008): Introduction to methods applied in hazard and risk analysis. In N Gebbeken, K Thoma (Eds.): "Bau-Protect" Buildings and Utilities Protection. Bilingual Proceedings. Berichte aus dem Konstruktiven Ingenieurbau Nummer 2008/02. 3. Workshop Bau-Protect, Buildings and Utilities Protection. Munich, 28.-29.10.2008. Neubiberg: Universität der Bundeswehr München.

Echterling, N. (2010). IEC 61508 Phasen und Methoden (presentation). Koblenz Symposium. Holzen.

Echterling, N. (2011). EMI report E 17/11

EDEN (2016): EU Project, End-user driven demo for CBRNE, 2013–2016, Grant agreement ID: 313077. Available online at https://cordis.europa.eu/project/id/313077; https://eden-securi tyfp7.eu/, checked on 9/27/2020.

Encounter (2015): EU Project Encounter, explosive neutralisation and mitigation countermeasures for IEDs in urban/civil environment, 2012–2015, Grant agreement ID: 285505 (2015-08-20). Available online at https://cordis.europa.eu/project/id/285505; www.encounter-fp7.eu/.

Ericson, C. (2005). Hazard Analysis Techniques for System Safety, Wiley.

Fehling-Kaschek, Mirjam; Faist, Katja; Miller, Natalie; Finger, Jörg; Häring, Ivo; Carli, Marco et al. (2019): A systematic tabular approach for risk and resilience assessment and Improvement in the telecommunication industry. In Michael Beer, Enrico Zio (Eds.): Proceedings of the 29th European Safety and Reliability Conference (ESREL 2019). ESREL. Hannover, Germany, 22–26 September 2019. European Safety and Reliability Association (ESRA). Singapore: Research Publishing Services, pp. 1312–1319. Available online at https://esrel2019.org/files/proceedings.zip.

Fehling-Kaschek, Mirjam; Miller, Natalie; Haab, Gael; Faist, Katja; Stolz, Alexander; Häring, Ivo et al. (2020): Risk and Resilience Assessment and Improvement in the Telecommunication Industry. In Piero Baraldi, Francesco Di Maio, Enrico Zio (Eds.): Proceedings of the 30th European Safety and Reliability Conference and the 15th Probabilistic Safety Assessment and Management Conference. ESREL2020 and PSAM15. European Safety and Reliability Aassociation (ESRA), International Association for Probabilistic Safety Assessment and Management (PSAM). Singapore: Research Publishing Services. Available online at https://www.rpsonline.com.sg/proceedings/esrel2020/pdf/3995.pdf, checked on 9/25/2020.

Fischer, K.; Riedel,W.; Häring, I.; Nieuwenhuijs, A.; Crabbe, S.; Trojaborg, S.; Hynes,W. (2012a): Vulnerability Identification and Resilience Enhancements of Urban Environments. In N. Aschenbruck, P. Martini, M. Meier, J. Tölle (Eds.): 7th Security Research Conference, Future Security 2012. Bonn.

Fischer, Kai; Riedel, Werner; Häring, Ivo; Nieuwenhuijs, Albert; Crabbe, Stephen; Trojaborg, Steen S. et al. (2012b): Vulnerability Identification and Resilience Enhancements of Urban Environments. In Nils Aschenbruck, Peter Martini, Michael Meier, Jens Tölle (Eds.): Future Security. 7th Security Research Conference, Future Security 2012, Bonn, Germany, September 4-6,

2012. Proceedings, vol. 318. Berlin, Heidelberg: Springer (Communications in Computer and Information Science, 318), pp. 176–179.

Fischer, K; Häring, I; Riedel, W (2015): Risk-based resilience quantification and improvement for urban areas. In Jürgen Beyerer, Andreas Meissner, Jürgen Geisler (Eds.): Security Research Conference. 10th Future Security; Berlin, September 15-17, 2015 Proceedings. Fraunhofer IOSB, Karlsruhe; Fraunhofer-Verbund Verteidigungs- und Sicherheitsforschung; Security Research Conference. Stuttgart: Fraunhofer Verlag, pp. 417–424. Available online at http://publica.fraunh ofer.de/eprints/urn_nbn_de_0011-n-3624247.pdf, checked on 9/25/2020.

Fischer, Kai; Hiermaier, Stefan; Riedel, Werner; Häring, Ivo (2018): Morphology Dependent Assessment of Resilience for Urban Areas. In *Sustainability* 10 (6), p. 1800. https://doi.org/ 10.3390/su10061800.

Fischer, Kai; Hiermaier, Stefan; Werner, Riedel; Häring, Ivo (2019): Statistical driven Vulnerability Assessment for the Resilience Quantification of Urban Areas. In: Proceedings of the 13-th International Confernence on Applications of Statistics and Probability in Civil Engineering. ICASP13. Seoul, South Korea, 26.-30.05.2019. Available online at http://s-space.snu.ac.kr/bitstr eam/10371/153259/1/26.pdf, checked on 9/25/2020.

Ganter, Sebastian; Srivastava, Kushal; Vogelbacher, Georg; Finger, Jörg; Vamanu, Bogdan; Kopustinskas, Vytis et al. (2020): Towards Risk and Resilience Quantification of Gas Networks based on Numerical Simulation and Statistical Event Assessment. In Piero Baraldi, Francesco Di Maio, Enrico Zio (Eds.): Proceedings of the 30th European Safety and Reliability Conference and the 15th Probabilistic Safety Assessment and Management Conference. ESREL2020 and PSAM15. European Safety and Reliability Aassociation (ESRA), International Association for Probabilistic Safety Assessment and Management (PSAM). Singapore: Research Publishing Services. Available online at https://www.rpsonline.com.sg/proceedings/esrel2020/html/3971. xml, checked on 9/25/2020.

Häring, I (2013a): Sorting enabling technologies for risk analysis and crisis management. In Bernhard Katzy, Ulrike Lechner (Eds.): Civilian Crisis Response Models (Dagstuhl Seminar 13041). 27 pages. With assistance of Marc Herbstritt, p. 71.

Häring, I. (2013b): Sorting enabling technologies for risk analysis and crisis management. In Bernhard Katzy, Ulrike Lechner (Eds.): Civilian Crisis Response Models (Dagstuhl Seminar 13041). 27 pages. With assistance of Marc Herbstritt, p. 71.

Häring, I. (2013c): Workshop research topic Resilient Systems. In Bernhard Katzy, Ulrike Lechner (Eds.): Civilian Crisis Response Models (Dagstuhl Seminar 13041). 27 pages. With assistance of Marc Herbstritt, pp. 86–87.

Häring, Ivo (2015): Risk analysis and management. Engineering resilience. Singapur: Springer

Häring, I. and B. Kanat (2013). "Standardization of best-practice safety analysis methods…" Draft.

Häring, I.; Schönherr, M.; Richter, C. (2009): Quantitative hazard and risk analysis for fragments of high-explosive shells in air. In *Reliability Engineering & System Safety* 94 (9), pp. 1461–1470. https://doi.org/10.1016/j.ress.2009.02.003.

Häring, Ivo; Ebenhöch, Stefan; Stolz, Alexander (2016): Quantifying resilience for resilience engineering of socio technical systems. In *Eur J Secur Res* (*European Journal for Security Research*) 1 (1), pp. 21–58. https://doi.org/10.1007/s41125-015-0001-x.

Häring, Ivo; Sansavini, Giovanni; Bellini, Emanuele; Martyn, Nick; Kovalenko, Tatyana; Kitsak, Maksim et al. (2017): Towards a generic resilience management, quantification and development process: general definitions, requirements, methods, techniques and measures, and case studies. In Igor Linkov, José Manuel Palma-Oliveira (Eds.): Resilience and risk. Methods and application in environment, cyber and social domains. NATO Advanced Research Workshop on Resilience-Based Approaches to Critical Infrastructure Safeguarding. Dordrecht: Springer (NATO science for peace and security series. Series C, Environmental security), pp. 21–80. Available online at https://www.springer.com/de/book/9789402411225.

Häring, Ivo; Ramin, Malte von; Stottmeister, Alexander; Schäfer, Johannes; Vogelbacher, Georg; Brombacher, Bernd et al. (2019): Validated 3D Spatial Stepwise Quantitative Hazard, Risk and Resilience Analysis and Management of Explosive Events in Urban Areas. In *Eur J Secur Res*

(European Journal for Security Research) 4 (1), pp. 93–129. https://doi.org/10.1007/s41125-018-0035-y.

Häring, Ivo; Ganter, Sebastian; Finger, Jörg; Srivastava, Kushal; Agrafioti, Evita; Fuggini, Clemente; Botella, Fabio (2020): Panarchy Process for Risk Control and Resilience Quantification and Improvement. In Piero Baraldi, Francesco Di Maio, Enrico Zio (Eds.): Proceedings of the 30th European Safety and Reliability Conference and the 15th Probabilistic Safety Assessment and Management Conference. ESREL2020 and PSAM15. European Safety and Reliability Aassociation (ESRA), International Association for Probabilistic Safety Assessment and Management (PSAM). Singapore: Research Publishing Services. Available online at https://www.rpsonline.com.sg/proceedings/esrel2020/pdf/4264.pdf, checked on 9/25/2020.

IEC 61508 (2010). Functional Safety of Electrical/Electronic/Programmable Electronic Safetyrelated Systems Edition 2.0 Geneva, International Electrotechnical Commission.

IEC 61508 S+ (2010): Functional Safety of Electrical/Electronic/Programmable Electronic Safetyrelated Systems Ed. 2. Geneva: International Electrotechnical Commission.

Jain, Aishvarya Kumar; Satsrisakul, Yupak; Fehling-Kaschek, Mirjam; Häring, Ivo; vanRest, Jeroen (2020): Towards Simulation of Dynamic Risk-Based Border Crossing Checkpoints. In Piero Baraldi, Francesco Di Maio, Enrico Zio (Eds.): Proceedings of the 30th European Safety and Reliability Conference and the 15th Probabilistic Safety Assessment and Management Conference. ESREL2020 and PSAM15. European Safety and Reliability Aassociation (ESRA), International Association for Probabilistic Safety Assessment and Management (PSAM). Singapore: Research Publishing Services. Available online at https://www.rpsonline.com.sg/proceedings/esrel2020/pdf/4000.pdf, checked on 9/25/2020.

Kaufman, J E; Häring, I (2011): Functional safety requirements for active protection systems from individual and collective risk criteria. In Christophe Berenguer, Antoine Grall, Carlos Guedes Soares (Eds.): Advances in Safety, Reliability and Risk Management. Esrel 2011. 1st ed. Baton Rouge: Chapman and Hall/CRC.

MABS 15: Med-Eng Systems Inc. Ottawa, Ontario, Canada.

MIL-STD-882D (2000). Standard Practice for System Safety. US Military Standard, Department of Defense, United States of Amerika.

Miller, N.; Fehling-Kaschek, M.; Haab, G.; Faist, K.; Stolz, A.; Häring, I. (2020): Resilience analysis and quantification for Critical Infrastructures. In John Soldatos, James Philpot, Gabriele Giunta (Eds.): Cyber-Physical Threat Intelligence for Critical Infrastructures Security: A Guide to Integrated Cyber-Physical Protection of Modern Critical Infrastructures. [S.l.]: Now Publishers, pp. 365–384.

Ministry of Defence (2007). Defence Standard 00-56, Issue 4. Safety management Requirements for Defence Systems.

OCTIKT (2020): Ein Organic-Computing basierter Ansatz zur Sicherstellung undVerbesserung der Resilienz in technischen und IKT-Systemen. German BMBF Project. Available online at https://projekt-octikt.fzi.de/, checked on 5/15/2020.

Renger, P; Siebold, U; Kaufmann, R; Häring, I (2015): Semi-formal static and dynamic modeling and categorization of airport checkpoints. In Tomasz Nowakowski (Ed.): Safety and reliability. Methodology and applications; [ESREL 2014 Conference, held in Wrocław, Poland. ESREL. London: CRC Press.

RESILENS (2018): EU project, Realising European resilience for critIcal infrastructure, 2015–2018, Grant agreement ID: 653260. Available online at https://cordis.europa.eu/project/id/653260; http://resilens.eu/, checked on 9/27/2020.

Riedel, W; Fischer, K; Stolz, A; Häring, I (2015): Modeling the vulnerability of urban areas to explosion scenarios. In M. G. Stewart, M. D. Netherton (Eds.): Design and analysis of protective structures. Proceedings of the 3rd International Conference on Protective Structures ICPS3: 3–6 February 2015 Newcastle Australia. International Conference on Protective Structures ICPS3. Callahan, NSW: Centre of Infrastructure Performance and Reliability, School of Engineering, University of Newcastle, pp. 469–478.

Riedel,W; Niwenhuijs, A; Fischer, K; Crabbe, S; Heyenes,W; Müllers, I et al. (2014): Quantifying urban risk and vulnerability – a tool suite of new methods for planners. In Klaus Thoma, IvoHäring, Tobias Leismann (Eds.): Proceedings/9th Future Security - Security Research Conference. Berlin, September 16–18, 2014. Fraunhofer-Verbund Verteidigungs- und Sicherheitsforschung; Future Security Research Conference; Future Security 2014. Stuttgart: Fraunhofer Verl., pp. 8–16.

Salhab, Rene G.; Häring, Ivo; Radtke, Frank K. F. (2011): Fragment launching conditions for risk analysis of explosion and impact scenarios. In G.G.S. Bérenguer and G. Soares (Ed.): ESREL 2011 Annual Conference. Troyes, France: Taylor and Francis Group, London.

Schäfer, Johannes; Rathjen, Sina (2014a): Risk acceptance criteria: Deliverable D6.5 for the D-BOX project (which has received funding from the European Union's Seventh Framework Programme for research, technological development and demonstration under Grant Agreement No 284996).

Schäfer, J.; Kopf, N.; Häring, I (2014b): Empirical risk analysis of humanitarian demining for characterization of hazard sources. In Klaus Thoma, Ivo Häring, Tobias Leismann (Eds.): Proceedings/9th Future Security - Security Research Conference. Berlin, September 16–18, 2014. Fraunhofer-Verbund Verteidigungs- und Sicherheitsforschung; Future Security Research Conference; Future Security 2014. Stuttgart: Fraunhofer Verl., pp. 598–602.

Siebold, U; Hasenstein, S; Finger, J; Häring, I (2015): Table-top urban risk and resilience management for football events. In Luca Podofillini, Bruno Sudret, Božidar Stojadinović, Enrico Zio, Wolfgang Kröger (Eds.): Safety and reliability of complex engineered systems. Proceedings of the 25th European Safety and Reliability Conference, ESREL 2015, Zürich, Switzerland, 7–10 September 2015. European Safety and Reliability Conference; ESREL 2015. Boca Raton, London, New York: CRC Press Taylor & Francis Group a Balkema book, pp. 3375–3382.

Stolz, Alexander; Roller, Christoph; Rinder, Tassilo; Siebold, Uli; Häring, Ivo (2017): Aktuelle Forschungsergebnisse zum baulichen Schutz für Großbauwerke und Risikomanagement für kritische Infrastrukturen und städtische Bereiche. In Stefan Hiermaier, Norbert Gebbeken, Michael Klaus, Alexander Stolz (Eds.): Gefährdung, dynamische Analyse und Schutzkonzepte für bauliche Strukturen. 7. Workshop Bau-Protect: Tagungsband. Ernst-Mach-Institut; Deutschland; Workshop Bau-Protect. Stuttgart: Fraunhofer Verlag, pp. 314–334.

Thielsch, P. (2012). Risikoanalysemethoden zur Festlegung der Gesamtssicherheitsanforderungen im Sinn der "IEC 61508 (Ed. 2)". Bachelor, Hochschule Furtwangen.

Tomforde, Sven; Gelhausen, Patrick; Gruhl, Christian; Häring, Ivo; Sick, Bernhard (2019): Explicit Consideration of Resilience in Organic Computing Design Processes. In: ARCS Workshop 2019 and 32nd International Conference on Architecture of Computing Systems. Joint Conference. Copenhagen, Denmark. Berlin, Germany: VDE Verlag GmbH, pp. 51–56.

VITRUV (2014): EU Project VITRUV, Vulnerability Identification Tools for Resilience Enhancements of Urban Environments, 2011–2014 (2015-08-20). Available online at https://cordis.eur opa.eu/project/id/261741, checked on 9/27/2020.

Vogelbacher, Georg; Häring, Ivo; Fischer, Kai; Riedel, Werner (2016): Empirical Susceptibility, Vulnerability and Risk Analysis for Resilience Enhancement of Urban Areas to Terrorist Events. In *Eur J Secur Res (European Journal for Security Research)* 1 (2), pp. 151–186. https://doi.org/10.1007/s41125-016-0009-x.

Vogelbacher, Georg; Finger, Jörg; Häring, Ivo (2019): Towards Visibility and Audibility Algorithms for Assessing Perceived Safety and Security in Public Areas Based on Digital 3d City Models. In Michael Beer, Enrico Zio (Eds.): Proceedings of the 29th European Safety and Reliability Conference (ESREL 2019). ESREL. Hannover, Germany, 22-26 September 2019. European Safety and Reliability Association (ESRA). Singapore: Research Publishing Services, pp. 1986–1993.

Voss, M; Häring, I; Fischer, K; Riedel, W; Siebold, U (2012): Susceptibility and vulnerability of urban buildings and infrastructure against terroristic threats from qualitative and quantitative risk analyses. In: 11th International Probabilistic Safety Assessment and Management Conference and the Annual European Safety and Reliability Conference 2012. (PSAM11 ESREL 2012); Helsinki, Finland, 25–29 June 2012. International Association for Probabilistic Safety Assessment and Management; European Safety and Reliability Association; International Probabilistic

Safety Assessment and Management Conference; PSAM; Annual European Safety and Reliability Conference; ESREL. Red Hook, NY: Curran, pp. 5757–5767.

XP-DITE (2017): Accelerated Checkpoint Design Integration Test and Evaluation. Booklet. Available online at https://www.xp-dite.eu/dissemination/, updated on 2017-2, checked on 9/25/2020.

Chapter 9
Reliability Prediction

9.1 Overview

The general goal of *reliability prediction* of a system is to estimate the probability that the system fails in a certain time interval as precisely as possible (Glose 2008).

For this, several standards based on different data and different models have been developed, see Sect. 9.10, which provides a tabular comparison scheme for tools. The concepts and definitions that are necessary to understand the concepts of those standards are explained in this chapter.

Section 9.2 defines the term reliability as a time-dependent probability expression of a system. Reliability is much more focused when compared to the broad concept of dependability which is a kind of quality concept for safety–critical and/or highly available systems.

In Sect. 9.3, a scheme is introduced which shows that reliability prediction complements the range of system analysis methods by providing frequency-type input, namely, system state analysis such as Markov analysis and system mode analyses like FMEA and FTA. It is emphasized that reliability prediction can provide input for any quantitative system analysis methodology, in particular, FMEA, FTA, and Markov analysis. As field data is more and more easily accessible due to increasing connectivity and intelligence, it is expected that scaling, updating, and even generation of empirical reliability prediction become feasible. Section 9.4 names some software tools that are used for reliability predictions.

Sections 9.5 to 9.8 introduce the terms failure, reliability, dependability, demand modes, failure density, and failure rate in the context of reliability predictions. Formulae are listed and connections between the quantities are drawn. A table is provided that shows the general connections and example connections for the exponential, the Weibull, and the lognormal distribution. The bathtub curve, a function of the failure rate in time, is introduced in Sect. 9.9.

Section 9.10 explains the general design of reliability prediction standards. Then, examples are given with focus on MIL-HDBK-217, a widely used standard, and

I. Häring, *Technical Safety, Reliability and Resilience*,
https://doi.org/10.1007/978-981-33-4272-9_9

FIDES, a more recent standard. The chapter ends with questions with answers in Sects. 9.12 and 9.13.

After introducing the necessary definitions, it will be shown that there are several standards for reliability predictions. They are based on additive or additive–multiplicative models. The standard MIL-HDBK-217 is still widely used, although it is not updated anymore. However, reliability growth factors are used to take account of improvement of technology and production. The modern standard 217-Plus is the successor of this standard. Companies have enriched the standard MIL-HDBK-217 with their own field data to get a more accurate standard for their field of work. The modern standard FIDES uses a different model than MIL-HDBK-217. Its goal is to give less conservative estimates of the probability of failure.

The chapter is mostly based on Meier and Häring (2010) and Ringwald and Häring (2010). Ringwald's main source is Birolini (2007). Meier's main sources are Bowles (1992), IEEE-Guide (2003), Birolini (2007), Glose (2008), Hoppe and Schwederski (2009).

9.2 Reliability and Dependability

Like in Chap. 7, we denote *reliability* with Rb instead of the common R to avoid confusions with R for risk.

From a qualitative point of view, reliability can be seen as the ability of a component to carry out the demanded functions under given conditions. From a quantitative perspective, reliability is the probability that there is no failure within a given time interval $(0, t]$ (Birolini 2007; Ringwald and Häring 2010).

In mathematical notation, reliability can be written as a function $Rb(t)$ with

$$Rb(t) = 1 - F(t) \quad \text{and} \quad \lim_{t \to \infty} Rb(t) = 0, \tag{9.1}$$

where $F(t) = P(\text{component fails in } (0, t])$ (Ringwald and Häring 2010).

Dependability is "the collective term used to describe the availability performance and its influencing factors: reliability performance, maintainability performance and maintenance support performance" (IEC 60050-191 1990).

Please note that, for instance, in German language, dependability and reliability both mean "Zuverlässigkeit." So, when deciding on the right English translation of "Zuverlässigkeit," one should keep in mind that reliability as defined above is only a part of the much broader concept of dependability (Glose 2008).

9.3 Embedding "Reliability Prediction" into the Range of System Analysis Methods

There are several methods to quantify the dependability of components or of the system. The scheme in Fig. 9.1 shows the common analyzing methods used in industry. It has similarities to the scheme in Fig. 4.4. but this time the focus is different. The focus is not on inductive and deductive methods. Instead the analyzing methods are divided into the three categories "system state analysis," "reliability prediction," and "failure modes analysis" which are explained in more detail in Sects. 9.3.1 to 9.3.3.

The main output of reliability prediction is failure models of single components. They typically depend on

- component type;
- technology used for the component;
- environmental stress the component is exposed to, e.g., temperature cycles, mechanical stress, radiation, humidity, salty air, and corrosive environment;
- internal stress the component is exposed to, e.g., electrical current loading, voltage loading, capacity loading, or percentage of capacity of the component used;
- Aging;
- Maturity level of component; and
- In some cases, experience of developers.

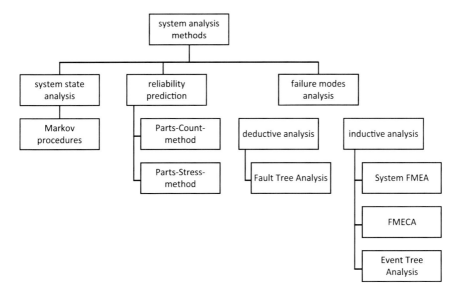

Fig. 9.1 The most common system analysis (failure rate analysis) methods, translated from Meier and Häring (2010)

9.3.1   Failure Modes Analysis

Failure modes analysis determines the potential types of failures that can occur. This can be done inductively by asking "Which failures can occur and which consequences do they have?" or deductively by regarding a potential failure and asking for the causes of this failure (Meier and Häring 2010).

Fault tree analysis (Chap. 6) and FMEA (Chap. 7) are examples for failure mode analysis methods.

9.3.2   Reliability Prediction

Reliability prediction computes the system's reliability using the failure rates of single elements and their interfaces. For this, several standards have been developed which describe methods to compute failure rates with regard to environmental influences (Meier and Häring 2010).

Reliability prediction will be introduced in this chapter.

9.3.3   System State Analysis

System state analysis computes the system's dependability and safety by applying Markov models. Markov models use state transition diagrams to display the system's dependability and safety performance. They are used to model the temporal behavior of the system where a system is understood as a composition of elements with only two states ("functioning" and "failed"). The system itself can be in many states, each of which is characterized by a certain combination of functioning and failed elements (EMPA 2009; Meier and Häring 2010).

System state analysis will not be treated in more detail here.

9.4   Software

There are several software tools for reliability predictions, for example, A.L.D. (RAM-Commander), Reliability Analysis Center (PRISM), and RELEX (Reliability Studio). Each of these tools uses different and in most cases identical methods and standards to determine component and system safety (Ringwald and Häring 2010).

Besides for (failure) predictions, the tools are also used to display the functional connection of the components in reliability block diagrams (RBD), for fault trees, and for event trees.

9.5 Failure

A *failure* occurs if a component cannot carry out its demanded function. Failure-free time, that is, the time where no failures occur, can usually be interpreted as a random variable. Here, it is assumed that the component is free of failures at $t = 0$. Besides their probability of occurrence, failures should be analyzed in terms of sort, cause, and consequences (Ringwald and Häring 2010).

In the following let

$$F(t) = P(\text{component fails in } (0, t]), \qquad (9.2)$$

that is, F(t) is defined as the probability that the component fails up to time t (Ringwald and Häring 2010).

9.6 Demand Modes for Safety Functions

When regarding the different parts of a system, one distinguishes between a high and a low *demand mode* of operation (IEC 61508 S+ 2010; Ringwald and Häring 2010).

In *low demand mode*, a system only uses its safety function once or very rarely (at most once a year) (Birolini 2007). For example, if the system is an airbag system, the safety function that the airbag deploys in a crash is in low demand mode. On the other hand, the safety function that the airbag does not deploy during normal driving is used constantly. It is hence in *high demand mode* (Ringwald and Häring 2010).

For a system in low demand mode, one regards the probability of failure on demand (PFD). A system in high demand mode is evaluated by the probability of failure per hour (PFH). With those, the *safety integrity levels* (SIL) are defined, see Table 8.11.

9.7 Failure Density

The *failure density* $f(t)$ is defined by

$$f(t) := \frac{d}{dt} F(t), \qquad (9.3)$$

where $F(t)$ is the probability that the component fails in $(0, t]$, see Sect. 9.5 (Glose 2008; Hansen 2009; Meier and Häring 2010).

The definition of $f(t)$ can be rewritten as

Table 9.1 Common density functions of reliability prediction (Glose 2008)

Distribution	Failure density $f(t)$	Failure probability $F(t) = \int\limits_{0}^{t} f(s)\,ds$	Failure rate for $t > 0$ $\lambda(t) = \frac{f(t)}{1-F(t)}$
Exponential	$f(t) = \begin{cases} \alpha \cdot e^{-\alpha t} & \text{for } t \geq 0, \\ 0 & \text{for } t < 0 \end{cases}$ $\alpha > 0$	$F(t) = \begin{cases} 1 - e^{-\alpha t} & \text{for } t \geq 0, \\ 0 & \text{for } t < 0 \end{cases}$ $\alpha > 0$	$\lambda(t) = \alpha$ $\alpha > 0$
Weibull	$f(t) = \begin{cases} \frac{b}{T}\left(\frac{t}{T}\right)^{b-1} e^{-\left(\frac{t}{T}\right)^{b}} & \text{for } t \geq 0, \\ 0 & \text{for } t < 0 \end{cases}$ $b,\ T > 0$	$F(t) = \begin{cases} 1 - e^{-\left(\frac{t}{T}\right)^{b}} & \text{for } t \geq 0, \\ 0 & \text{for } t < 0 \end{cases}$ $b,\ T > 0$	$\lambda(t) = \frac{b}{T}\left(\frac{t}{T}\right)^{b-1}$ $b,\ T > 0$
Log-normal	$f(t) = \begin{cases} \frac{1}{t\sigma\sqrt{2\pi}} e^{-\frac{(\ln t - \mu)^2}{2\sigma^2}} & \text{for } t > 0, \\ 0 & \text{for } t \leq 0 \end{cases}$ $\mu \in \mathbb{R},\ \sigma > 0$	$F(t) = \begin{cases} \frac{1}{\sigma\sqrt{2\pi}} \int\limits_{0}^{t} \frac{1}{s} e^{-\frac{(\ln s - \mu)^2}{2\sigma^2}}\,ds & \text{for } t > 0, \\ 0 & \text{for } t \leq 0 \end{cases}$ $\mu \in \mathbb{R},\ \sigma > 0$	$\lambda(t) = \frac{e^{-\frac{(\ln t - \mu)^2}{2\sigma^2}}}{t\sigma\sqrt{2\pi}(1-F(t))}$ $\mu \in \mathbb{R},\ \sigma > 0$

$$
\begin{aligned}
f(t) &:= \frac{d}{dt} F(t) \\
&= \lim_{\delta t \to 0} \frac{P(\text{failure in } (0, t + \delta t]) - P(\text{failure in } (0, t])}{\delta t} \\
&= \lim_{\delta t \to 0} \frac{P((\text{failure in } (t, t + \delta t]) \cap (\text{no failure in } (0, t]))}{\delta t}
\end{aligned} \tag{9.4}
$$

which will be needed in Sect. 9.8. The failure density can be interpreted as the rate of change of the failure probability.

Common density functions with the corresponding failure probabilities and failure rates are listed in Table 9.1. An overview of mathematical definitions is given in Table 9.2.

9.8 Failure Rate

The *failure rate* $\lambda(t)$, given that the system is free of failure at time 0, is defined by the following formula (Leibniz-Rechenzentrum 2004):

$$
\lambda(t) := \lim_{\delta t \to 0} \frac{P(\text{failure in } (t, t + \delta t] \mid \text{no failure in } (0, t])}{\delta t}. \tag{9.5}
$$

This can be written as

Table 9.2 Overview of mathematical definitions for reliability predictions (Meyna and Pauli 2003)

English term	Mathematical definition/notation/formulas
Actual time of failure T	T is a random variable
Survival probability $Rb(t)$	$Rb(t) = P(T > t)$ with: $Rb(u) \leq Rb(t)$ for $u > t$, $Rb(0) = 1$, $\lim_{t \to \infty} Rb(t) = 0$, $Rb(t) = \int_t^\infty f(s)ds$
Failure probability $F(t)$	$F(t) = P(T \leq t)$ with: $F(u) \geq F(t)$ for $u > t$, $F(0) = 0$, $\lim_{t \to \infty} F(t) = 1$, $F(t) = \int_0^t f(s)ds$
Event density $f(t)$	$f(t) = \frac{d}{dt} F(t)$ with: $f(t) = 0$ for $t < 0$, $f(t) \geq 0$ for $t \geq 0$, $\int_0^\infty f(t)dt = 1$
Hazard rate/failure rate $\lambda(t)$	$\lambda(t) = \frac{f(t)}{Rb(t)} = \frac{f(t)}{1-F(t)}$
Mean time between failures (MTBF), Average working time of system/expected value $E(T)$	$E(T) = \int_0^\infty Rb(t)dt$

$$
\begin{aligned}
\lambda(t) &:= \lim_{\delta t \to 0} \frac{P(\text{failure in } (t, t + \delta t] \mid \text{no failure in } (0, t])}{\delta t} \\
&= \lim_{\delta t \to 0} \frac{P((\text{failure in } (t, t + \delta t]) \cap (\text{no failure in } (0, t]))}{\delta t \cdot P(\text{no failure in } (0, t])} \\
&= \frac{f(t)}{P(\text{no failure in } (0, t])} \\
&= \frac{f(t)}{1 - F(t)},
\end{aligned}
\tag{9.6}
$$

where we use (9.4) in the third equation. Informally spoken, the failure rate is the rate with which a failure occurs at time t if it is known that the failure has not yet occurred until time t.

As unit of measurement, one typically uses FIT (failures in time). The definitions of FIT vary in the different standards. For example, in MIL-HDBK-217 1 FIT $= 10^{-6}/h$ while in FIDES 1 FIT $= 10^{-9}/h$ (Ringwald and Häring 2010).

Example (Ringwald and Häring 2010) Let the probability that a certain component fails during a 30-year-long storage period be 10^{-8}. This yields a failure rate of $\frac{10^{-8}}{30a} \approx 10^{-8} \cdot \frac{1}{30} \cdot \frac{1}{365} \cdot \frac{1}{24h} \approx 3.81 \cdot 10^{-14}/h = 3.81 \cdot 10^{-5}$ FIT with the definition from FIDES.

If the failure rate $\lambda(t)$ is constant, that is, $\lambda(t) \equiv \lambda$, its reciprocal value $1/\lambda$ is the mean in time to the next failure (MTTF) (Glose 2008).

9.9 Bathtub Curve

Starting point to determine the failure rate is the observed failures, see Fig. 9.2. Those are composed of the early "infant mortality" failures, the constant random failures, and the wear out failures. Adding the relating curves (failures as a function of time) yields the *bathtub curve* (Meyna and Pauli 2003; Ringwald and Häring 2010).

The bathtub curve can, for example, be seen as the sum of two failure rate curves belonging to two Weibull distributions with parameters $b_1 = 0.5, b_2 = 5$, and $T = 1$ (Glose 2008), see Fig. 9.3.

One typically divides the bathtub curve into three parts (Birolini 1991):

– Part 1: The part where the curve $\lambda(t)$ decreases quickly. A high number of early failures can be caused by material weaknesses, quality fluctuations during the production process, or application errors (dimensioning, testing, operation).
– Part 2: The part of the curve where $\lambda(t)$ is almost constant.
– Part 3: The part of the curve where $\lambda(t)$ increases due to wear out failures.

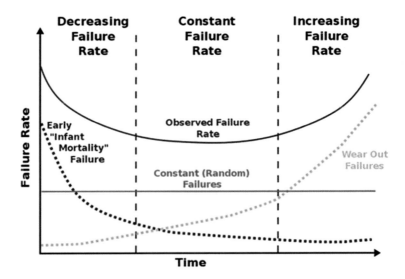

Fig. 9.2 Schematic of Bathtub curve, see, e.g., Rajarshi and Rajarshi (1988)

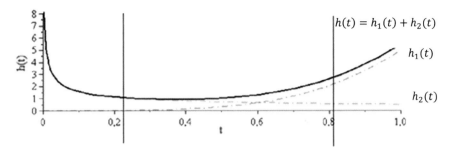

Fig. 9.3 Example bathtub curve as the sum of two Weibull distributions (Glose 2008)

Most standards only refer to the second phase and assume a constant failure rate λ, only some standards, for example, 217-Plus, use a factor to regard the early failures (Glose 2008; Meier and Häring 2010). In practical applications, it has turned out that the assumption of a constant failure rate is realistic for a lot of systems and makes the computations easier (Birolini 2007). On the other hand, there also are several studies, for example, (Wong 1989; Wong 1995; Jones and Hayes 2001), that criticize this assumption (Glose 2008).

9.10   Standards

Besides the logical and structural arrangement of the components, the failure rates of the single components are essential in order to determine the failure rate of the whole system. *Standards* are used to compute those failure rates. The standards are described in handbooks which introduce generic failure rates that have been gained through field data and simulation tests. Furthermore, the standards establish the functional connections between failure rates and influencing factors such as temperature or humidity of the surroundings (Meier and Häring 2010).

This section gives an overview of the standards' general design and then lists examples that are used in the industry.

9.10.1   General Design

Standards are normally described in handbooks which contain one or more of the following prediction methods (IEEE-Guide 2003):

- Tables with constant component-specific failure rates, for operation under load as well as for operation in idle.
- Factors for the environmental influences to compute the constant failure rates for operation under load as well as for operation in idle.

- Factors for converting between failure rates for operation under load and operation in idle.

One distinguishes between multiplicative and multiplicative–additive models. Most standards have adapted to the multiplicative model of MIL-HDBK-217, see Sect. 9.10.2. For those multiplicative models, the formula to compute the failure rate of a component is

$$\lambda_P = f(\lambda_G, \pi_1, \pi_2, \ldots, \pi_i) = \lambda_G \, \pi_1 \, \pi_2 \, \ldots \, \pi_i, \qquad (9.7)$$

where λ_P is the predicted failure rate of the component, λ_G is the generic failure rate, and π_k with $1 \leq k \leq i$ are adjustment factors (Meier and Häring 2010).

Modern standards such as FIDES or 217-Plus use multiplicative–additive models (Meier and Häring 2010). For these, the formulas to compute the failure rate contain additions and multiplications. The formulae for the different multiplicative–additive models do not have a generic form in common like the multiplicative models with the formula in (9.7). The formulae have more complex structures than (9.7) and have to be regarded individually for the different standards. Here, this will be done for FIDES in Sect. 9.10.11.

9.10.2   MIL-HDBK-217

The standard *MIL-HDBK-217* was developed by the DoD (Department of Defense, USA) and RADC (Rome Air Development Center, today: Rome Laboratory) in 1960 and is the most well-known standard for reliability predictions of electronic components. Although the failure rates are based on field data from 1980 to 1990 and the handbook has not been updated since 1995, it is still the most widely used standard for reliability predictions (Hoppe and Schwederski 2009; Meier and Häring 2010; Ringwald and Häring 2010).

MIL-HDBK-217 uses multiplicative models to determine failure rates of electrical, electronic, and electromechanical components considering parameters such as environmental conditions, quality, loads, etc. It consists of two prediction methods, "Parts Stress" and "Parts Count," which will be explained in the following two subsections (Ringwald and Häring 2010).

9.10.2.1   Parts Stress

The *Parts Stress method* is used toward the end of the design process when a detailed parts list (which includes component loads, environmental conditions, and arrangement of the components) is available. The fundamental equation for the computation of a component's predicted failure rate λ_p is

$$\lambda_p = \lambda_b \cdot \pi_E \cdot \pi_Q \cdot \ldots \tag{9.8}$$

with generic failure rate λ_b (representing the influence of electrical and temperature loads on the component), the environmental adjustment factor π_E, and the quality adjustment factor π_Q.

Environmental conditions are composed into groups that relate to the typical military operation areas (MIL-HDBK-217E: Table 5.1.1-3). These three general factors (λ_b, π_E, and π_Q) are contained in every component model (Department Of Defense 1986). Further π-factors depend on the type of component (resistor, capacitor, …) and regard additional component-typical stress factors (MIL-HDBK-217E: Table 5.1.1-4) (Ringwald and Häring 2010).

9.10.2.2 Parts Count

The *Parts Count method* is used to estimate the failure rate in early design phases. It is especially useful if there is not enough information available to cay out the Parts Stress method (Ringwald and Häring 2010).

The Parts Count method categorizes components by types (resistor, capacitor, …) and uses a predefined failure rate for each type of component (see the tables in Sect. 5.2 of MIL-HDBK-217E (Department Of Defense 1986)). The total failure rate of the system is computed by

$$\lambda_{total} = \sum_{i=1}^{n} N_i \ (\lambda_G \cdot \pi_Q)_i \tag{9.9}$$

where λ_{total} is the failure rate of the whole system, λ_G is the type-specific failure rate of the ith component, π_Q is the component's quality of the ith component, N_i says how many of the ith component are incorporated into the system, and n is the number of different components used in the system. If a component is produced for less than 2 years or if there are drastic changes in the design or in the process, an additional factor π_L, the so-called learning factor or reliability growth factor, is multiplied with $\lambda_G \cdot \pi_Q$ for this component (Ringwald and Häring 2010). The learning factor becomes smaller as the product becomes older and more reliable. Experimental data (for example, by analyzing the number of failures before and after the change of a component) and statistic tools are used to determine the learning factor.

In (9.9), adding up all failure rates means that a serial linking of all components is assumed. In general, this overestimates the actual failure rate. Hence, the Parts Count method normally yields higher and less precise failure rates than the Parts Stress method (Glose 2008).

9.10.3   SN29500 (Siemens)

The Siemens standard can be used for electronic and electromechanical components (Meier and Häring 2010).

The failure rates are given in the measuring unit "failures per $10^6 h$" and only refer to the operation under loads (Hoppe and Schwederski 2009). They are determined by Siemens test and field data as well as external sources such as MIL-HDBK-217. The structure of the models is similar to the one of MIL-HDBK-217 (Bowles 1992) but the adjustment factors only cover influences by electrical and thermal loads (ECSS 2006; Meier and Häring 2010).

9.10.4   Telcordia

The Telcordia standard was originally known as Bell Laboratories Bellcore standard for reliability predictions for usual electronic components (Ringwald and Häring 2010).

Telcorida TR-332 Issue 6 and SR-332 Issue 1 are used as economical standards in telecommunication. Telcordia comprises three methods to predict a product's reliability: Parts Count, Parts Count with additional use of laboratory data, and reliability prediction using field data. Those methods have additive and multiplicative parts (Ringwald and Häring 2010).

9.10.5   217-Plus

217-Plus is an advancement of MIL-HDBK-217F. The adjustment factors have been improved by new data and the model has been restructured for a more precise computation (Ringwald and Häring 2010).

9.10.6   NSWC

The NSWC Standard 98/LE1 ("Handbook of Reliability Prediction Procedures for Mechanical Equipment") is a standard used by the American Naval Surface Warfare Center for reliability predictions of mechanical components in the military (Ringwald and Häring 2010).

9.10.7 IEC TR 62380

IEC TR 62380 was developed by the French company CNET (Centre National d'Etudes des Télécommunications, today: France Télécom (Glose 2008)). Modeling of thermal behavior and systems at rest is used to take different environmental conditions into account (Meier and Häring 2010).

The same data that was used by CNET was also used by two other telecommunication companies, BT (British Telecom) and Italtel (Italy), for their own standards HRD5 (Handbook of Reliability Data) and IRPH 2003 (Italtel Reliability Prediction Handbooks), respectively (Glose 2008).

9.10.8 IEEE Gold Book (IEEE STD 493-1997)

The IEEE Gold Book is an IEEE recommendation for the design of dependable industrial and commercial electricity supply systems. It includes data for commercial voltage distribution systems (Meier and Häring 2010).

9.10.9 SAE (PREL 5.0)

Due to bad predictions of MIL-HDBK-217 with respect to failure rates in the field, the Society of Automotive Engineers (SAE), USA, enriched the standard with their own data in 1983 (Denson 1998; Glose 2008).

9.10.10 GJB/Z 299B

GJB/Z 299B was developed by the Chinese military and comprises, like the MIL-HDBK-217, the Parts Stress and the Parts Count methods. It was translated to English in 2001. Sometimes it is also referred to as "China 299B" (Vintr 2007).

9.10.11 FIDES

The modern standard *FIDES* was developed by a consortium of French companies which was led by the French Department of Defense. It estimates the reliability, with focus on rough environmental conditions (for example, in defense, aeronautical, industrial economic, or transport systems) (Meier and Häring 2010).

FIDES is based on field data, tests of component manufacturers, and existing reliability prediction methods. It wants to give realistic (not pessimistic or conservative) estimates. It is also possible to use the standard on sub-assemblies instead of only on individual components. This makes it possible to use the standard early in the development process (Meier and Häring 2010).

In addition to the failure considered in MIL-HDBK-217, FIDES also regards insufficiencies in the development process, failures in the production process, wrong assumptions, and induced failures. The influences are divided into three categories: technology, process, and use. The category "technology" covers the component itself and its installation, "process" includes all procedures from development to decommissioning, and "use" considers restrictions in use by the manufacturer and the qualification of the user. Software failures and improper use are not considered (Meier and Häring 2010).

The total failure rate of the system is computed by

$$\lambda_{total} = \sum_{i=1}^{n} \left(\lambda_{physical} \pi_{process} \pi_{PM} \right)_i, \tag{9.10}$$

where $\lambda_{physical}$ is the ith component's physical failure rate; $\pi_{process}$ is a measure for the quality of the development, production, and operating process; and π_{PM} (PM = parts manufacturing) is a quality evaluation of the component and the manufacturer (Meier and Häring 2010).

The physical failure rate $\lambda_{physical}$ is determined by the equation

$$\lambda_{physical} = \pi_{induced} \cdot \sum_{k=1}^{m} (\lambda_0 \cdot \pi_{acceleration})_k, \tag{9.11}$$

where $\pi_{induced}$ is a factor for mechanical, thermal, and electrical overloading; λ_0 are the constant base failure rates; and $\pi_{acceleration}$ are thermal, thermally cyclic, electrical, or mechanical acceleration factors. For each acceleration factor, there is exactly one base failure rate (Glose 2008; Meier and Häring 2010).

9.11 Comparison of Standards and Software Tools

For comparison of the coverage of standards along with available software implementations, one can compare in tabular form for each standard the following criteria (Glose 2008):

- Scope of application of standard.
- Last edition.
- Expectation of updates.

- Software tools implementing standards taking account of their maturity, scientific, and operational acceptance.
- Units accessible by software tools.
- Number and scope of predefined environmental conditions.
- Coverage of environmental conditions of particular importance for application, e.g., translational and high-dynamic acceleration, temperature extremes, high altitudes, or weather conditions.
- Types of models used for reliability prediction.
- Consideration of system operation profile.
- Consideration of thermal cycles.
- (Explicit) Consideration of temperature (increase) during operation.
- (Explicit) Consideration of loss of solder connection.
- (Explicit) Consideration of induced faults.
- (Explicit) Consideration of year of production.
- Consideration of development process (experience with similar systems).
- Consideration of software safety.
- (Explicit) Consideration of burn-in failures.
- Consideration of failure of non-operating components.
- Consideration of existing similar systems.
- Consideration of test results.

9.12 Questions

(1) For which system analysis methods can you use reliability prediction?
(2) Which reliability prediction standards can be distinguished?
(3) How are Pi factors determined?
(4) Which information is typically not yet available in case when applying parts count analysis?
(5) What are reliability growth factors?
(6) What is the difference between multiplicative and additive reliability prediction.
(7) Which of the following expressions are true?

    (a) It is assumed that the component is free of failures at $t = 0$,
    (b) $F(t) = P(\text{component fails in } (0, t])$,
    (c) $F(t) = P(\text{component fails after time } t)$,
    (d) $F(t) = P(\text{component does not fail in } (0, t])$,
    (e) $Rb(t) = P(\text{component fails in } (0, t])$,
    (f) $Rb(t) = P(\text{component fails after time } t)$, and
    (g) $Rb(t) = P(\text{component does not fail in } (0, t])$.

(8) Which of the following expressions are true?

    (a) $\lim_{t \to \infty} Rb(t) = 1$,

(b) $\lim\limits_{t \to \infty} F(t) = 0$,

(c) $\lim\limits_{t \to \infty} Rb(t) = 0$,

(d) $\lim\limits_{t \to \infty} F(t) = 1$, and

(e) $F(t) + Rb(t) = 1$.

(9) Which of the following is true?

　　(a) Dependability includes reliability.
　　(b) Dependability is a part of reliability.
　　(c) Dependability and reliability mean the same.

(10) Which of the following is true?

　　(a) $\lambda(t) = \frac{f(t)}{F(t)}$,

　　(b) $\lambda(t) = \frac{f(t)}{Rb(t)}$.

(11) Name the three types of system analysis methods and explain them in your own words.

(12) Name the three parts of the bathtub curve. What is the usual assumption?

(13) What is the generic formula for multiplicative reliability prediction models?

(14) When was the last time MIL-HDBK-217 was updated?

　　(a) 1990,
　　(b) 1995,
　　(c) 2000, and
　　(d) 2005.

(15) When should the parts stress method be used, when is the parts count method better?

(16) Name two standards that are based on MIL-HDBK-217.

(17) Regard an exponential distribution with $\alpha = 2$.

　　(a) Compute $f(1)$, $F(1)$, and $\lambda(1)$.

　　(b) Show that $\lim\limits_{t \to \infty} Rb(t) = 0$, $\lim\limits_{t \to \infty} F(t) = 1$, and $\int\limits_{0}^{\infty} f(t)dt = 1$ hold.

　　(c) What is $E(T)$?

(18) Regard the bathtub curve as the sum of two Weibull distributions as in Fig. 9.3 with $b_1 = 0.5$, $b_2 = 5$, and $T = 1$.

For which time t is $\lambda(t)$ minimal? Which value does it take?

9.13 Answers

(1) For all quantitative methods, e.g., FMEDA, FTA, and Markov models.

(2) Classification options include: multiplicative versus additive, standards that consider the development process or not, type and number of environmental conditions (external stress) and internal loading (operational stress) considered, and resolution of technologies.

(3) For example, from field data using multi-regression analysis.

(4) The internal loading of components.

(5) Factors that are multiplied with the failure base rate of traditional standards that consider technology maturity production improvement effects.

(6) Compare the equations discussed in Sect. 9.10. See Sect. 9.10.1

(7) (a), (b), (g), see Sects. 9.5 and 9.2.

(8) (c), (d), (e), see Sect. 9.2.

(9) (a), see Sect. 9.2.

(10) (b), see Sect. 9.8.

(11) See Sects. 9.3.1 to 9.3.3.

(12) See Sect. 9.9.

(13) See Sect. 9.10.

(14) (b), see Sect. 9.10.2.

(15) Parts count is better earlier in the development process because it needs less information. Parts stress should be chosen if enough information is available to carry it out because it is more precise.

(16) SN 29500 (Siemens), 217-Plus, SAE (PREL 5.0), (GJB/Z 299B), see Sect. 9.10.

(17) See Table 9.1:

(a) $f(1) = 2e^{-2}$, $F(1) = 1 - e^{-2}$, $\lambda(1) = 2$.

(b) $\lim\limits_{t\to\infty} F(t) = \lim\limits_{t\to\infty} \left(1 - e^{-2t}\right) = 1 - 0 = 1$, $\lim\limits_{t\to\infty} Rb(t) =$

$\lim\limits_{t\to\infty} (1 - F(t)) = 1 - 1 = 0$, $\int\limits_0^\infty f(t)dt = \int\limits_0^t f(t)dt + \int\limits_t^\infty f(t)dt = F(t) +$

$Rb(t) = 1$.

(c) $E(T) = \int\limits_0^\infty Rb(t)dt = \int\limits_0^\infty e^{-2t}dt = \left[-\tfrac{1}{2}e^{-2t}\right]_0^\infty = \tfrac{1}{2} = \tfrac{1}{\lambda}$.

(18) $\lambda(t) = \lambda_1(t) + \lambda_2(t) = 0.5 \cdot t^{-0.5} + 5 \cdot t^4$

Necessary condition: $0 = -0.25 \cdot t^{-1.5} + 20 \cdot t^3$. This yields $t = \left(\tfrac{1}{80}\right)^{\frac{2}{9}} \approx 0.378$. $\lambda''(0.378) \approx 12.8 > 0$, hence it is a minimum.
The corresponding failure rate is $\lambda(0.378) \approx 0.915$.

References

Birolini, A. (1991). Qualität und Zuverlässigkeit technischer Systeme. Berlin, Heidelberg, Springer.

Birolini, A. (2007). Reliability Engineering: Theory and Practice, Springer Berlin.

Bowles, J. B. (1992). "A Survey of Reliability-Prediction Procedures For Microelectronic Devices." IEEE Transactions on Reliability **41**(1): 2–12.

Denson, W. (1998). "The History of Reliability Prediction." IEEE Transactions on Reliability **47**(3-SP): 321–328.

Department Of Defense (1986). Military Handbook - Reliability Prediction of Electronic Equipment (MIL-HDBK-217E), U.S. Government Painting Office.

ECSS (2006). Space Product assurance - Components reliability data sources and their use, Noordwijk, The Netherlands, ESA Publications Division.

EMPA. (2009). "Markoff-Modelle." Retrieved from 2012-04-02.

Glose, D (2008): Zuverlässigkeitsvorhersage für elektronische Komponenten unter mechanischer Belastung, Reliability predicition of electronic components for mechanical stress environment. Hochschule Kempten, University of Applied Sciences; Fraunhofer EMI, Efringen-Kirchen.

Hansen, P. (2009). Entwicklung eines energetischen Sanierungsmodells für den europäischen Wohngebäudesektor unter dem Aspekt der Erstellung von Szenarien für Energieund CO2-Einsparpotenziale bis 2030, Forschungszentrum Jülich.

Hoppe, W. and P. Schwederski (2009). Comparison of 4 Reliability Prediction Approaches for realistic failure rates of electronic parts required for Safety & Reliability Analysis. Zuverlässigkeit und Entwurf - 3. GMM/GI/ITG-Fachtagung. Stuttgart, Germany.

IEC 60050-191 (1990). Internationales Elektrotechnisches Wörterbuch. Genf, International Electrotechnical Commission.

IEC 61508 S+ (2010). Functional Safety of Electrical/Electronic/Programmable Electronic Safety-related Systems Ed. 2 Geneva, International Electrotechnical Commission.

IEEE-Guide (2003). "IEEE Guide for selecting and using reliability prediction based on IEEE 1413."

Jones, J. and J. Hayes (2001). "Estimation of System Reliability Using a "Non-Constant Failure Rate" Model." IEEE Transaction on Reliability **50**(3): 286–288.

Leibniz-Rechenzentrum. (2004). "Verweildaueranalyse, auch: Verlaufsdatenanalyse, Ereignisanalyse, Survival-Analyse (engl.: Survival Analysis, Analysis of Failure Times, Event History Analysis)." Retrieved 2012-04-11, from http://www.lrz.de/~wlm/ilm_v8.htm.

Meier, S. and I. Häring (2010). Zuverlässigkeitsanalyse der an der induktiven Datenübertragung beteiligten Schaltungsteile eines eingebetteten Systems. Praxisarbeit DHBW Lörrach/EMI-Bericht, Fraunhofer Ernst-Mach-Institut.

Meyna, A. and B. Pauli (2003). Taschenbuch der Zuverlässigkeits- und Sicherheitstechnik; Quantitative Bewertungsverfahren. München, Hanser.

Rajarshi, Sujata; Rajarshi, M. B. (1988): Bathtub distributions: a review. In *Communications in Statistics - Theory and Methods* 17 (8), pp. 2597–2621. https://doi.org/10.1080/036109288088 29761.

Ringwald, M. and I. Häring (2010). Zuverlässigkeitsanalyse der Elektronik einer Sensorplatine. Praxisarbeit DHBW Lörrach/ EMI-Bericht, Fraunhofer Ernst-Mach-Institut.

Vintr, M. (2007). Tools for Reliability Prediction of Systems, Brno University of Technology.

Wong, K. L. (1989). "The Roller-Coaster Curve is in." Quality and Reliability Engineering int. **5**: 29–36.

Wong, K. L. (1995). "A New Framework For Part Failure-Rate Prediction Models." IEEE Transaction on Reliability **44**(1): 139–146.

Chapter 10
Models for Hardware and Software Development Processes

10.1 Overview

System development processes historically always have been driven by scarcity and availability of resources and knowledge. In the software domain, examples include the costs of computation time and memory when compared to labor programming costs or the accessibility of external computation power and memory to local systems. In the hardware domain, examples include the costs of real-time testing when compared with emulated or virtual testing and verification.

A further example is the at least intellectually, however often not emotionally and not lived, broadly accepted insight that the increasing complexity of systems and hence systems development requires the explicit implementation of actively obeyed suitably structured development processes. In this context, a challenging requirement is to take account of tacit or implicit rules, decision-making processes as well as technical processes.

This includes the insight that flexibility and efficiency of processes can be driven by less as well as by more process requirements. On the one hand, loose requirement regarding the consistency of specifications may lead to long-term inconsistencies and expensive efficiency loss at later stages of project development. On the other hand, rather strong formal requirements regarding system specifications may prevent from fast prototype development for early customer and end user feedback, however much increase efficiency at later project stages. Of course, both statements can also be reversed depending on the system development context. Hence, it is important to balance process requirements with flexibility expectations.

In particular, the initially intended functionality and structure of systems, i.e., the requirements that are most relevant for successful system validation, are typically much more accessible at early stages of a development project when compared to later stages. Therefore, system development processes should try to mine this knowledge and install it in the development process as early as possible. The term validation is used in this chapter in the sense that it assesses whether the overall system delivers what customers and end users really expect in realistic applications, foreseeable similar applications as well as potential and accessible misuse.

Taking these challenges into account, this chapter introduces different types of software and hardware development processes. However, instead of following the doctrines of any of the plethora of existing system development processes and their communities, first selected generic properties of any development process are identified.

The chapter starts with a list of attributes or characteristics of development processes in Sect. 10.2 that may also occur in combination for a single development process. Afterward, a few well-known examples of models for software and hardware development processes are introduced and presented in typical process schemes, see Sect. 10.3.

The development model properties incremental, iterative, and linear are identified as key properties of development processes using software development as sample case. They can be employed in a balanced way in systematic system development processes to make systems safer. The introduced different software development models can be classified in traditional models and agile models. The attributes introduced are helpful to show that also agile development methods use very similar approaches as classical methods, however adapted to much increased computational resources, the work force expectations, and the need for fast progress validation and monitoring in modern economic contexts.

In the next chapter, we take a closer look at the Safety Life Cycle of the IEC 61508. The life of a system is divided into 16 phases which we will introduce.

The development processes presented here may incorporate activities for achieving a high level of safety of systems. However, achieving safety and security is not the main aim of standard development processes of systems. As system development processes cover the development of systems, they can be used to structure the overall life cycle of a system from "cradle to grave," i.e., from the first idea to the final dismantling, reuse, or recycling, especially the first phases, see Sect. 3.3.

In Chap. 11, we will see that safety development life cycles can be understood as special cases or extensions of development life cycles.

10.2 Properties of the Software Development Models

The models that will be introduced in Sect. 10.3 describe different types of development processes. Three attributes or properties of such models are listed in this section.

10.2.1 *Incremental Versus Big Bang Development*

"*Incremental* development is a staging and scheduling strategy in which various parts of the system are developed at different times or rates, and integrated as they are completed. The alternative strategy to incremental development is to develop the entire system with a 'big bang' integration at the end" (Cockburn 2008).

In case of incremental development, the development steps may be very small. Typically, for each step, all the steps are performed as if a major development step is performed, including verification and validation of the development.

Example A computer program that is completely written before it is compiled for the first time is not developed incrementally. An incremental development would occur in several steps with compiling different parts of the program at different times.

10.2.2 *Iterative Development*

"*Iterative* development is a rework scheduling strategy in which time is set aside to revise and improve parts of the system. The alternative strategy to iterative development is to plan to get everything right the first time" (Cockburn 2008).

Example A computer program that is written once and sold right away is not iteratively developed. A program where different functions are improved several times after having been tested is iteratively developed.

10.2.3 *Linear Development (Without or With Steps Back)*

A development process is called *linear* if it is not possible to go back to previous steps of the development, that is, if the development steps are executed in a predefined order and conclude in a final result.

10.2.4 *Agile Software Development*

Agile software development is based on incremental and iterative development. The term was introduced in 2001 and is based on the "Manifesto for Agile Software Development" (Beck et al. 2001) and its values.

"We are uncovering better ways of developing software by doing it and helping others do it. Through this work we have come to value:

Individuals and interactions over processes and tools. Working software over comprehensive documentation. Customer collaboration over contract negotiation. Responding to change over following a plan.

That is, while there is value in the items on the right, we value the items on the left more" (Beck et al. 2001).

Iterations are short time intervals (normally a few weeks) and involve a team. After each iteration the results are shown to the client/decision-maker.

An example of agile software development is Scrum, see Sect. 10.3.5.

10.3 Example Development Models

10.3.1 Waterfall Model

The *Waterfall Model* organizes the development process into phases, which have to be validated in mile stone meetings. Only if one phase is completed, the development process can proceed to the next phase, see Fig. 10.1. Steps to the previous phase can only be done in the extended Waterfall Model, see Fig. 10.2. See the references Kramer (2018) and Ruparelia (2010) for some recent applications and discussions.

Each phase has an output in the form of documents or reports. The requirements listed in these documents have to be precise enough to be used for surveying the implementation.

"The sequential phases in Waterfall model are:

– **Requirement Gathering and analysis**: All possible requirements of the system to be developed are captured in this phase and documented in a requirement specification doc.
– **System Design**: The requirement specifications from first phase are studied in this phase and system design is prepared. System Design helps in specifying hardware and system requirements and also helps in defining overall system architecture.

Fig. 10.1 Sample pure waterfall system development model without option to step back

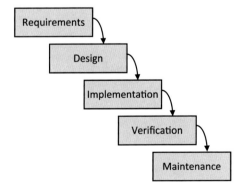

Fig. 10.2 Sample extended waterfall system development model with the option to step back (dashed arrow) to the latest development step only

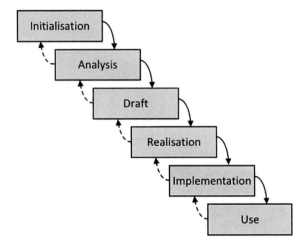

- **Implementation**: With inputs from system design, the system is first developed in small programs called units, which are integrated in the next phase. Each unit is developed and tested for its functionality which is referred to as Unit Testing.
- **Integration and Testing**: All the units developed in the implementation phase are integrated into a system after testing of each unit. Post integration the entire system is tested for any faults and failures.
- **Deployment of system**: Once the functional and non-functional testing is done, the product is deployed in the customer environment or released into the market.
- **Maintenance**: There are some issues which come up in the client environment. To fix those issues patches are released. Also, to enhance the product some better versions are released. Maintenance is done to deliver these changes in the customer environment.

All these phases are cascaded to each other in which progress is seen as flowing steadily downward (like a waterfall) through the phases. The next phase is started only after the defined set of goals are achieved for previous phase and it is signed off, so the name 'Waterfall Model'. In this model phases do not overlap" (Tutorialspoint 2013),

"Advantages of waterfall model:

- Simple and easy to understand and use.
- Easy to manage due to the rigidity of the model—each phase has specific deliverables and a review process.
- Phases are processed and completed one at a time.
- Works well for smaller projects where requirements are very well understood.

Disadvantages of waterfall model:

- Once an application is in the testing stage, it is very difficult to go back and change something that was not well-thought out in the concept stage.

– No working software is produced until late during the life cycle.
– High amounts of risk and uncertainty.
– Not a good model for complex and object-oriented projects.
– Poor model for long and ongoing projects.
– Not suitable for the projects where requirements are at a moderate to high risk of changing.

When to use the waterfall model:

– Requirements are very well known, clear and fixed.
– Product definition is stable.
– Technology is understood.
– There are no ambiguous requirements.
– Ample resources with required expertise are available freely.
– The project is short" (ISTQB Guide 2013).

10.3.2   Spiral Model

The *Spiral Model* is an iterative software development process which is divided into four segments. These four segments are (see also Fig. 10.3) (Boehm 1988)

(1)   Determine objectives,
(2)   Identify and resolve risks,
(3)   Development and Test, and
(4)   Plan the next iteration.

The process starts in phase (1) and proceeds through the four phases. After the fourth phase, the first iteration is completed and the system has been improved. The process starts again in phase (1), only this time with the improved system. In this way, the system is improved with every iteration. This continues until the system meets the expectations. For example, a model with four iterations could divide into

(a)   Developing a concept for the production of the desired system,
(b)   Setting up a list of requirements that have to be regarded during the development of the system,
(c)   Developing and validating a first design of the system, and
(d)   Developing and implementing a final design of the system that then passes the final inspection.

In Kaduri (2012), the four steps are shortly described as follows:
"**Planning**: In this phase, the new system requirements are gathered and defined after a comprehensive system study of the various business processes. This usually involves interviewing internal and external users, preparation of detailed flow diagrams showing the process or processes for which going to be developed, the inputs and outputs in terms of how the data is to be recorded/entered and the form in which the results are to be presented.

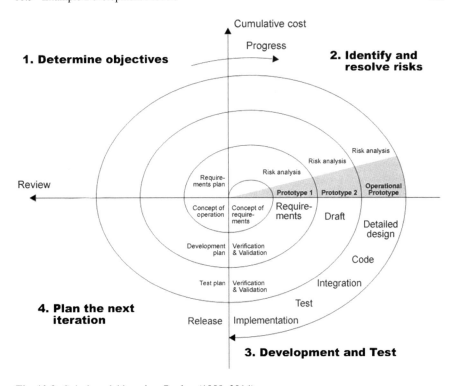

Fig. 10.3 Spiral model based on Boehm (1988, 2014)

Risk Analysis Phase: In this phase to identify risk and alternate solutions, a process is followed. The process includes addressing the factors that may risk the successful completion of the entire project development including alternative strategies and constraints. The issues pertaining to the possibility of the development if not meeting, for example, user requirements, reporting requirements or the capability of the development team or the compatibility and functionality of the hardware with software. To undertake development, the Risk analysis and suggested solutions to mitigate and eliminate the Risks would thus become a part of the finalized strategy. A prototype is produced at the end of the risk analysis phase.

Engineering Phase: In this phase, software is produced along with coding and testing at the end of the phase. After preparation of prototype tested against benchmarks based on customer expectations and evaluated risks to verify the various aspects of the development. Until customer satisfaction is achieved before development of the next phase of the product, refinements and rectifications of the prototype are undertaken.

Evaluation Phase: In this phase, the final system is thoroughly evaluated and tested based on the refined prototype. Evaluation phase allows the customer to evaluate the output of the project to date before the project continues to the next spiral.

Routine maintenance is carried out on a continuing basis to prevent large scale failures and to minimize downtime" (Manual Testing Guide 2013).

Rouse (2007, 2020) subdivides the four steps further and defines the following nine steps:

1. "The new system requirements are defined in as much detail as possible. This usually involves interviewing a number of users representing all the external or internal users and other aspects of the existing system.
2. A preliminary design is created for the new system.
3. A first prototype of the new system is constructed from the preliminary design. This is usually a scaled-down system, and represents an approximation of the characteristics of the final product.
4. A second prototype is evolved by a fourfold procedure: (1) evaluating the first prototype in terms of its strengths, weaknesses, and risks; (2) defining the requirements of the second prototype; (3) planning and designing the second prototype; (4) constructing and testing the second prototype.
5. At the customer's option, the entire project can be aborted if the risk is deemed too great. Risk factors might involve development cost overruns, operating-cost miscalculation, or any other factor that could, in the customer's judgment, result in a less-than-satisfactory final product.
6. The existing prototype is evaluated in the same manner as was the previous prototype, and, if necessary, another prototype is developed from it according to the fourfold procedure outlined above.
7. The preceding steps are iterated until the customer is satisfied that the refined prototype represents the final product desired.
8. The final system is constructed, based on the refined prototype.
9. The final system is thoroughly evaluated and tested. Routine maintenance is carried out on a continuing basis to prevent large-scale failures and to minimize downtime" (Rouse 2007).

Main observations include the following:

- Classical development steps are only part of the model.
- As important dimension, the cost levels are considered in terms of distance from the origin.
- The model allows for incremental as well as more waterfall-like development processes.
- Early forms of validation take place in early phases for retargeting development aims.
- Risk and target of development assessment is conducted at all system development levels, in particular, in idea, concept, design, and implementation phase.
- Schematic puts great emphasis on activities that are not pure engineering and development but focuses on refinement of targets, analysis, verification, and validation.

10.3.3 V-Model

The *V-Model*, see Figs. 10.4 and 10.5, is a graphical model in form of the letter "V" that represents the steps of the development of a system in its life cycle. It describes the activities that have to be executed and results that have to be produced during the development process.

The left side of the V covers the identification of requirements and creation of system specifications. They completely determine the system. The specifications are verified during the right-hand side steps. Already, in these steps, the verification/validation plans are developed.

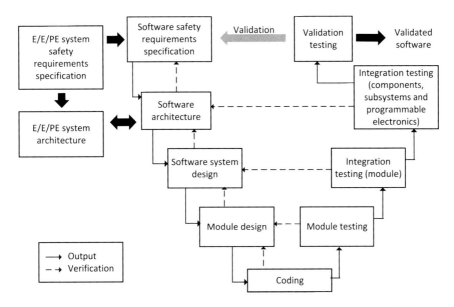

Fig. 10.4 V-Model from IEC 61508-3 (IEC 61508 2010). Copyright © 2010 IEC Geneva, Switzerland. www.iec.ch

Fig. 10.5 Less detailed example of a V-Model according to V-Model XT, see Angermeier et al. (2006)

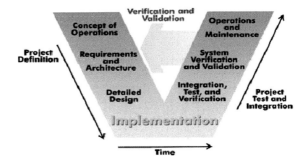

At the bottom of the V, the implementation at the highest level of resolution takes place.

The right side of the "V" represents integration of parts and their verification (Forsberg and Mooz 1991, 1998; Forsberg et al. 2005; International Council On Systems Engineering 2007; SEOR 2007).

Since the verification/validation activities at higher level of resolution (toward the bottom of the V) are more technical, often only at highest level of resolution, i.e., at system level, the term validation is used.

The left side of the "V" is comparable to the Waterfall Model. It describes the development of the system from general requirements to a precise draft. The right side of the "V" describes the verification and validation process. Here the process starts with the more precise testing, the verification. It is checked whether the actual requirements for the system have been fulfilled. The validation which follows the verification is the more general test. Here it is tested whether the general concept has been implemented correctly ("Does the system really do what was planned?").

More recent discussions and modifications of the V model include Bröhl and Dröschel (1995), Dröschel (1998), Clark (2009), Mathur and Malik (2010), Eigner et al. (2017) and Angermeier et al. (2006).

There also exists a widely used extension to software project management called V-Model XT, see Angermeier et al. (2006), Friedrich (2009), for which by now a rich experience of application exists.

10.3.4   Rational Unified Process (RUP)

The rational unified process (RUP) is an iterative, incremental, and UML-supported software development process and process framework (Rational 1998; Kruchten 2007; Kopal 2016). It consists of four phases: concept phase, design phase, development phase, and implementation (launch) phase.

An example of migration from RUP to agile development is described in Shafiee et al. (2020). A critical paper regarding RUP is given by Hesse (2001).

The approach is not described here in more detail because it is based on UML which will be covered in Chap. 13. It can be considered as a typical best practice approach taking in addition account of software semi-formal modeling at high abstraction level.

10.3.5   Scrum

Scrum is an agile software development method suitable for small groups as well as development teams (Rising and Janoff 2000; Beck et al. 2001; Schwaber and Beedle 2002). In contrast to the Waterfall or Spiral Model, scrum consists of many iterations, so-called sprints. The main idea of this agile software development is to identify and maintain a prioritized list of development aims (backlogs) that should be achieved, each of which typically lasts a week.

It is then decided jointly by the team, orchestrated and supervised by the Scrum master, which development target is conducted next at the beginning of each sprint. Scrum meetings are designed in an interactive personal but formal fashion as well as feedback and verification and validation sessions at the end of each sprint.

Using the terminology introduced in Sect. 10.2, the Scrum software development process is incremental in terms of software aims and developments achieved within each sprint. And it is iterative since sprints are repeated every week and agile, i.e., adaptive to findings during the development process as well as customer feedback to (sets of sprint) results.

The application of the Scrum method to embedded software is described in Takahira et al. (2012). Scrum application relative to other development processes is described in Kuhrmann et al. (2017).

An application of Scrum to the functional safety domain is described in Wang et al. (2017), which also refers to earlier preparatory conceptual work. A straightforward option is to define sprints for certain phases of the V-Models, in particular, the implementation work given the functional safety system specification. Thus, Scrum ideas can be implemented to a wide degree even for functional safety applications and when fulfilling the requirements of IEC 61508.

Main observations for the Scrum approach include the following:

- Allows to include end users in early phases of software development to include early first validation feedback.
- Use of specific terminology including

  - Product backlog: reservoir of desired product features and properties,
  - Sprint backlog: product features selected for implementation,
  - Scrum (meeting, "crowding together"): daily, weekly monthly,
  - Product owner (responsible person): end user,
  - (Working) Team, and
  - Scrum master (process responsible person).

- Covers also the design of (pro) active interactions of developers.
- Assigns by now well-established roles, including verification and validation.
- Scrum is typically well accepted and asked for by developers also in the safety domain since short feedback loops increase, in particular, (perceived) self-efficiency, overall project communication, and fast adoption to new overall project targets.
- Scrum corresponds to technological options to launch early Beta versions of software applications for early participative user feedback.
- Approach can be used for efficient improvement of existing applications by improving and adding features.

10.4   Questions

(1) Which properties (classifications) of system development processes do you know?
(2) Which classical models for the software development process do you know? Describe the different models in your own words.
(3) Assign to the development models the characteristics Sect. 10.2.
(4) What are the main advantages of the V-Model in the context of safety-relevant and safety–critical systems?

10.5   Answers

(1) See Sect. 10.2.
(2) See Sect. 10.3 focusing on extended Waterfall, V-Model, and Scrum.
(3) Examples: The pure Waterfall Model is linear. The Spiral Model is incremental and iterative. The "V"-Model is a modification of an extended Waterfall Model but not linear anymore due to the verification and validation specifications that start in early phases. Also, early verification and validation processes might result in changing designs and parts of the system in early development phases hinting at an (implicit) iterative nature of the V-Model.
(4) Advantages include the following: (i) implementation requires to define validation and verification criteria in early project development phases thus supporting later development phases; (ii) well established in the normalization landscape in many application domains; (iii) can be combined with agile development approaches, e.g., Scrum; (iv) combinations with hardware development exist, in particular, for embedded systems; and (v) steps backward and iterations are allowed.

References

Angermeier, D; Bartelt, C; Bauer, O; Benken, G; Bergner, K; Birowicz, U et al. (2006): Das V-Model XT. With assistance of Informationstechnikzentrum Bund (ITZBund). Edited by Verein zur Weiterentwicklung des V-Modell XT e.V. (Weit e.V.). Munich, Germany (Version 2.3). Available online at https://www.cio.bund.de/Web/DE/Architekturen-und-Standards/V-Modell-XT/vmodell_xt_node.html, checked on 9/21/2020.
Beck, Kent; Beedle, Mike; van Bennekum, Arie; Cockburn, Alistair; Cunningham, Ward; Fowler, Martin et al. (2001): Manifesto for Agile Software Development (2013-12-11). Available online at http://agilemanifesto.org/.
Boehm, B. W. (1988): A spiral model of software development and enhancement. In *Computer* 21 (5), pp. 61–72. https://doi.org/10.1109/2.59.
Boehm, Barry W. (2014): A Spiral Model of Software Development and Enhancement. In Ronald M. Baecker (Ed.): Readings in Human-Computer Interaction. Toward the Year 2000. 1. Aufl. s.l.: Elsevier Reference Monographs (Interactive Technologies), pp. 281–292.

Bröhl, A.-P.; Dröschel, W. (1995): Das V-Modell. München: Oldenburg Verlag.

Clark, John O. (2009): System of Systems Engineering and Family of Systems Engineering from a standards, V-Model, and Dual-V Model perspective. In: 3rd Annual IEEE Systems Conference, 2009. 23–26 March 2009, Marriott Vancouver Pinnacle Hotel, Vancouver, British Columbia, Canada. Piscataway, NJ: IEEE, pp. 381–387.

Cockburn, A. (2008). "Using Both Incremental and Iterative Development." STSC CrossTalk (USAF Software Technology Support Center) **21**(5): 27–30.

Dröschel, Wolfgang (1998): Inkrementelle und objektorientierte Vorgehensweisen mit dem V-Modell: Oldenburg.

Eigner, Martin; Dickopf, Thomas; Apostolov, Hristo (2017): The Evolution of the V-Model: From VDI 2206 to a System Engineering Based Approach for Developing Cybertronic Systems. In José Ríos, Alain Bernard, Abdelaziz Bouras, Sebti Foufou (Eds.): Product lifecycle management and the industry of the future. 14th IFIP WG 5.1 International Conference, PLM 2017: Seville, Spain, July 10-12, 2017: revised selected papers, vol. 517. International Federation for Information Processing; International Conference on Product Lifecycle Management; IFIP International Conference on Product Lifecycle Management; IFIP PLM; PLM. Cham: Springer (IFIP advances in information and communication technology, 517), pp. 382–393.

Forsberg, K. and H. Mooz (1991). The Relationship of Systems Engineering to the Project Cycle. First Annual Symposium of the National Council On Systems Engineering (NCOSE).

Forsberg, K. and H. Mooz. (1998). "System Engineering for Faster, Cheaper, Better." Retrieved 2003-04-20, from http://web.archive.org/web/20030420130303/, http://www.incose.org/sfbac/welcome/fcb-csm.pdf.

Forsberg, K., H. Mooz and H. Cotterman (2005). Visualizing Project Management. New York, John Wiley and Sons.

Friedrich, Jan (2009): Das V-Modell® XT: Für Projektleiter und QS-Verantwortliche kompakt und übersichtlich. s.l.: Springer Berlin Heidelberg.

Hesse, Wolfgang (2001): Dinosaur Meets Archaeopteryx? Seven Theses on Rational's Unified Process (RUP). Universtiy Marburg. Available online at https://www.mathematik.uni-marburg.de/~hesse/papers/Hes_01b.pdf.

IEC 61508 (2010): Functional Safety of Electrical/Electronic/Programmable Electronic Safety-related Systems Edition 2.0. Geneva: International Electrotechnical Commission.

International Council On Systems Engineering (2007). Systems Engineering Handbook Version 3.1: 3.3–3.8.

ISTQB Guide. (2013). "What is Waterfall model- advantages, disadvantages and when to use it?" Retrieved 2013-12-09, from http://istqbexamcertification.com/what-is-waterfall-model-advantages-disadvantages-and-when-to-use-it/.

Kaduri, Raja Rajo (2012): Manual testing material. ASAP D.D. (P) Ltd. Visakhapatnam. Available online at https://de.scribd.com/document/88814531/ASAP-D-D-P-Ltd-Manual-Testing-Material, checked on 9/21/2020.

Kopal, Nils (2016): Rational Unified Process. Universtiy Kassel. Available online at https://arxiv.org/pdf/1609.07350.pdf, checked on 9/21/2020.

Kramer, Mitch (2018): Best Practices in Systems Development Lifecycle: An Analyses Based on the Waterfall Model. In *Review of Business & Finance Studies*, 9 (1), pp. 77–84. Available online at https://ssrn.com/abstract=3131958, checked on 9/21/2020.

Kruchten, Philippe (2007): The rational unified process. An introduction. 3. ed., 7. printing. Upper Saddler River, NJ: Addison-Wesley (The Addison-Wesley object technology series).

Kuhrmann, Marco; Hanser, Eckhart; Prause, Christian R.; Diebold, Philipp; Münch, Jürgen; Tell, Paolo et al. (2017): Hybrid software and system development in practice: waterfall, scrum, and beyond. In A. Special Interest Group on Software C.M. Engineering (Ed.): Proceedings of the 2017 International Conference on Software and System Process. [Place of publication not identified]: ACM, pp. 30–39.

Manual Testing Guide. (2013). "Spiral Model." Retrieved 2013-12-11, from http://qtp-scripts.com/page/13/.

Mathur, Sonali; Malik, Shaily (2010): Advancements in the V-Model. In *IJCA* 1 (12), pp. 30–35. https://doi.org/10.5120/266-425.

Rational (1998): Rational Unified Process. Best Practices for Software Development Teams. Cupertino, USA (Rational Software White Paper TP026B, Rev 11/01). Available online at https://www.ibm.com/developerworks/rational/library/content/03July/1000/1251/1251_best practices_TP026B.pdf, checked on 9/21/2020.

Rising, L.; Janoff, N. S. (2000): The Scrum software development process for small teams. In *IEEE Softw.* 17 (4), pp. 26–32. https://doi.org/10.1109/52.854065.

Rouse, Margaret (2007): Spiral model (spiral lifecycle model) (2013-12-10). Available online at http://searchsoftwarequality.techtarget.com/definition/spiral-model.

Rouse, Margaret (2020): Spiral model. TechTarget, SearchSoftware Quality. Available online at https://searchsoftwarequality.techtarget.com/definition/spiral-model, updated on 2019-08, checked on 9/25/2020.

Ruparelia, Nayan B. (2010): Software development lifecycle models. In *SIGSOFT Softw. Eng. Notes* 35 (3), pp. 8–13. https://doi.org/10.1145/1764810.1764814.

Schwaber, Ken; Beedle, Mike (2002): Agile software development with Scrum. Upper Saddle River, NJ: Prentice Hall (Series in agile software development).

SEOR. (2007). "The SE VEE." Retrieved 2007-05-26, from http://www.gmu.edu/departments/seor/ insert/robot/robot2.html.

Shafiee, Sara; Wautelet, Yves; Hvam, Lars; Sandrin, Enrico; Forza, Cipriano (2020): Scrum versus Rational Unified Process in facing the main challenges of product configuration systems development. In *Journal of Systems and Software* 170, p. 110732. https://doi.org/10.1016/j.jss.2020. 110732.

Takahira, R Y; Laraia, L R; Dias, F A; Yu, A S; Nascimento, P T; Camargo, A S (2012): Scrum and Embedded Software development for the automotive industry. In Dundar F. Kocaoglu (Ed.): Proceedings of PICMET '12: technology management for emerging technologies (PICMET), 2012. Conference; July 29, 2012–Aug. 2, 2012, Vancouver, BC, Canada. Portland International Center for Management of Engineering and Technology; PICMET conference. Piscataway, NJ: IEEE, pp. 2664–2672.

Tutorialspoint. (2013). "SDLC Waterfall Model." Retrieved 2013-12-09, from http://www.tutorials point.com/sdlc/sdlc_waterfall_model.htm.

Wang, Yang; Ramadani, Jasmin; Wagner, Stefan (2017): An Exploratory Study on Applying a Scrum Development Process for Safety-Critical Systems. In Michael Felderer, Daniel Méndez Fernández, Burak Turhan, Marcos Kalinowski, Federica Sarro, Dietmar Winkler (Eds.): Product-Focused Software Process Improvement. 18th International Conference, PROFES 2017, Innsbruck, Austria, November 29-December 1, 2017, Proceedings, vol. 10611. Cham: Springer International Publishing (Lecture Notes in Computer Science, 10611), pp. 324–340.

Chapter 11
The Standard IEC 61508 and Its Safety Life Cycle

11.1 Overview

The international standard IEC 61508 "Functional Safety of Electrical/Electronic/Programmable Electronic (eepe, E/E/PE) Safety-related Systems" describes a standardized procedure to develop safe systems. The standard claims to be applicable to all systems that contain safety-related E/E/PE systems and where the failure of these safety-related systems causes danger for humans or the environment (Redmill 1998a, b, 2000a, b; Larisch, Hänle et al. 2008a, b). It can also be applied ex post to legacy systems (Schmidt and Häring 2007).

Since it is a generic norm (level A norm), it has to be adapted to the area of work where it should be used. This can be conducted using existing application standards (level B norms) or, if they are not (yet) available, by informed application of the generic standard to a new domain. The informed application of functional safety to new technological domains is key for successful development of new technologies.

All the system analysis methods from this book as described in Chap. 2, Sect. 3.16, Chaps. 5 to 9, Chaps. 13, 14, and 16 can be used when applying the standard to a system.

This chapter describes the standard IEC 61508 and explains the procedure of developing a safe system in case that EEPE systems significantly contribute to the generation of safety. Section 11.2 gives a brief summary of how the standard was developed and the updating history. Section 11.3 lists the names of the different parts of the standard and a scheme to describe the general structure.

Sections 11.4 to 11.6 remind of definitions and concepts from IEC 61508 that were already introduced and add selected further terms from the standard to this list, in particular, equipment under control (EUC), safety-related system, and a formal definition of safety function.

Sections 11.7 and 11.8 first introduce the safety life cycle and then its different 16 phases by giving a summary of the objectives, inputs, and outputs of each phase. For each phase, also sample methods are given that are also introduced in the present book that can be used to fulfill the requirements of the standard.

© The Author(s), under exclusive license to Springer Nature Singapore Pte Ltd. 2021
I. Häring, *Technical Safety, Reliability and Resilience*,
https://doi.org/10.1007/978-981-33-4272-9_11

The realization phase of the safety life cycle comprises the development of hardware (see Sect. 11.8.8.1) and software (see Sect. 11.8.8.2) and its respective mutual integration. An example is given of recommendations of methods for software for a selected SIL level.

The chapter is mostly based on the standard itself (IEC 61508 2010) and on (Larisch, Hänle et al. 2008a, b; Hänle 2007).

11.2   History of the Standard

The "International Electrotechnical Commission" (IEC) started the development of a standard for the safety of software-based systems in the 1980s. Two separate teams worked independently of each other on the software and hardware aspects and published two documents in 1992 which were unified to the draft standard IEC 1508 in 1995 (Redmill 1998a, b; Larisch, Hänle et al. 2008a, b).

The standard IEC 61508 resulted from this draft in 1998. Parts were modified in 2000 (Larisch, Hänle et al. 2008a, b). The second version of the norm was published in 2010.

11.3   Structure of the Standard

The standard IEC 61508 is divided into seven parts:

Part 1: General requirements,

Part 2: Requirements for electrical/electronic/programmable electronic safety-related systems,

Part 3: Software requirements,

Part 4: Definitions and abbreviations,

Part 5: Examples of methods for the determination of safety integrity levels,

Part 6: Guidelines on the application of IEC 61508-2 and IEC 61508-3, and.

Part 7: Overview of techniques and measures.

Parts 1–4 are normative and Parts 5–7 are informative (Liggesmeyer and Rombach 2005). Figure 11.1 shows the structure of the standard and illustrates the relations between the different parts.

The first part of the standard treats the basic safety life cycle. It describes the complete development of the safety requirements of an E/E/PE system. Parts 2 and 3 are closely linked. They contain the procedure for the hardware and software development. The fourth part consists of definitions and abbreviations that are used in the standard.

Parts 5–7 are informative and mostly contain further explanations and examples. They indirectly contain further requirements in form of tables that are necessary for Parts 2 and 3 (Larisch, Hänle et al. 2008a, b).

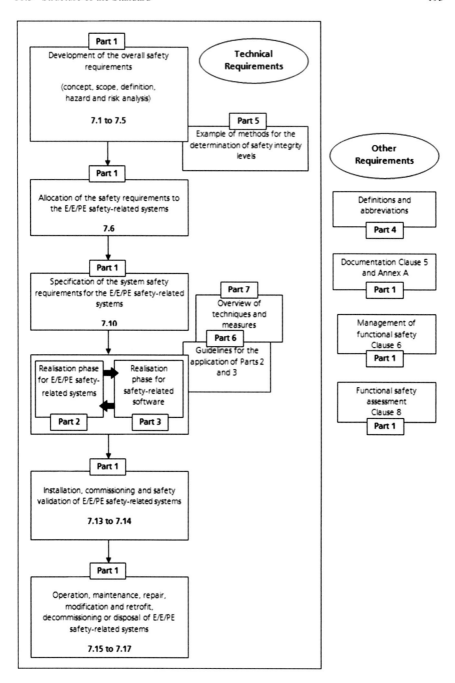

Fig. 11.1 Structure of the norm IEC 61508, graphic from (IEC 61508 2010) with optic modifications by Larisch as in (Larisch, Hänle et al. 2008a, b). Copyright © 2010 IEC Geneva, Switzerland. www.iec.ch

Table 11.1 Concepts and terms from IEC 61508 in previous chapters

Concept/term	Sections
Safety	Section 3.7
Risk minimization	Section 3.8
Safety function	Section 3.13
Safety integrity level	Section 3.13
Safety life cycle	Section 3.14
Dangerous failure	Section 7.9.5
Risk graph and risk parameter	Section 8.10.2
Computation of SIL	Section 3.13
Demand mode: continuous and on demand	Sections 3.13 and 9.6
V-Model	Section 10.3.3

Some examples of further literature on the IEC 61508 are the papers (Redmill 1998a, b; Bell 2005) and the books (Smith and Simpson 2004; Börcsök 2006).

11.4   Reminder

Several concepts from the IEC 61508 have already been mentioned or introduced throughout the course. They are listed in Table 11.1.

11.5   Definitions

Part 4 of the IEC 61508 lists definitions and abbreviations. We include the definitions here that are most important for the context of this course.

EUC (Equipment Under Control): "equipment, machinery, apparatus or plant used for manufacturing, process, transportation, medical or other activities" (IEC 61508 2010).

Element: "part of a subsystem comprising a single component or any group of components that performs one or more element safety functions" (IEC 61508 2010).

Remark: In Version 2 of IEC 61508, the term "element" is often used where it said "component" in Version 1.

Tolerable risk: "risk which is accepted in a given context based on the current values of society" (IEC 61508 2010).

EUC risk: "risk arising from the EUC or its interaction with the EUC control system" (IEC 61508 2010).

The relationship between tolerable risk and EUC risk was already shown in Fig. 3.2.

Safety-related system: "designated system that both.

- implements the required safety functions necessary to achieve or maintain a safe state for the EUC; and
- is intended to achieve, on its own or with other E/E/PE safety-related systems and other risk reduction measures, the necessary safety integrity for the required safety functions" (IEC 61508 2010).

Harm: "physical injury or damage to the health of people or damage to property or the environment" (IEC 61508 2010).

Hazardous situation: "circumstance in which people, property or the environment are exposed to one or more hazards" (IEC 61508 2010).

Hazard event: "event that may result in harm" (IEC 61508 2010).

Hazard: "potential source of harm" (IEC 61508 2010).

Hardware failure tolerance (HFT) $HFT = n$ means that system still operates if any n components are removed, where $n = 0, 1, 2, \cdots$.

Level of complexity of safety-related element (Level A/B) Level A means that all failure modes are known and are considered in analyses. Level B means that not all failure modes are known or cannot be considered explicitly in analysis. In particular, for Level A systems, all failure modes and their effect on the safety-related EEPE system are known as well as on overall system.

Diagnostic coverage (DC) DC is the percentage of failure that can be detected of safety-related EEPE systems. Often instead also the safe failure fraction (SFF) is used. See Sect. 7.9.5 on FMEDA for quantitative definitions of both terms.

Standard contains tables where maximum achievable SIL is provided given HFT, Level A/B, and DC. High SILs are only achievable for high HFT, Level A, and high DC.

11.6 Safety Function

A *safety function* is a protection mechanism of an E/E/PE (electrical/electronic/programmable electronic) system. According to IEC 61508, it is a "function to be implemented by an E/E/PE safety-related system or other risk reduction measures, that is intended to achieve or maintain a safe state for the EUC, in respect of a specific hazardous event" (IEC 61508 2010).

IEC 61508 distinguishes between qualitative and quantitative descriptions of safety functions. A qualitative description of a safety function would, for example, be (Hänle 2007)

$$A = \{\text{Blade stops when protection is removed within less than 0.1 sec.}\}. \quad (11.1)$$

A related quantitative description would be that the probability of failure of the safety function per request is smaller than a given value, for example,

$$P(A) < 1 \cdot 10^{-3}.$$

11.7 Safety Life Cycle

The standard IEC 61508 has a systematically structured safety concept, the *Safety Life Cycle*, see Fig. 11.2. It has to be kept in mind that this figure is only a scheme. In reality, the phases will not be sequentially gone through once but several steps will be repeated iteratively (Larisch, Hänle et al. 2008a, b).

The Safety Life Cycle helps to do all relevant measures in a systematic order to determine and achieve the safety integrity level (SIL) for a safety-related E/E/PE system. The safety life cycle consists of 16 phases which describe the complete life of a product, from the concept over installation and maintenance to decommissioning (Larisch, Hänle et al. 2008a, b).

The safety life cycle also considers safety-related systems from other technologies and external systems for risk reduction but they are not regarded in the rest of the standard IEC 61508. To proceed from one phase to the next, the present phase has to be completed. Work in later phases can result in the necessity of changes in earlier phases so that they have to be revised. The steps backward are not shown in Fig. 11.2 but iterations are allowed and sometimes necessary as mentioned before (Larisch, Hänle et al. 2008a, b).

During the first five phases of the Safety Life Cycle, the hazard and risk analysis are executed and based on those the safety requirements analysis. In Phase 9 and 10, the software and the hardware are developed and realized. Parallel to this phase, the

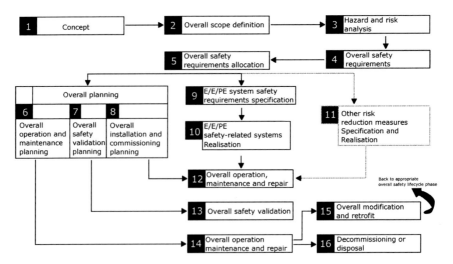

Fig. 11.2 Safety Life Cycle of IEC 61508 (IEC 61508 S+ 2010). Copyright © 2010 IEC Geneva, Switzerland. www.iec.ch

overall validation, installation, and operation are planned in Phases 6 to 8. The plans from these phases are executed in Phases 12 to 14. After changes to the system in Phase 15, one has to go back to the related phase in the Safety Life Cycle. Phase 16 describes the end of the Safety Life Cycle with the commissioning or disposal of the system.

Phase 11 is not treated in detail in the IEC 61508 (Larisch, Hänle et al. 2008a, b). However, its implications have to be considered throughout the application of the standard. A typical example is that a mechanical device significantly reduces the risk of a system, thus also lowering the overall SIL requirements for the remaining safety functions that are realized using hardware and software.

The standard IEC 61508 gives good and elaborate support for the later phases concerning the realization of the system. For the first phases up to the allocation of safety functions, experiences have shown that users often miss precise information from the standard (Larisch, Hänle et al. 2008a, b). The system analysis in Phases 1 to 3 which is necessary to determine the risks is not described in much detail, especially the evaluation of the acceptance of the risks is only shown exemplarily.

11.8   More Detailed Description of Some Phases

In the following sections, some of the 16 phases will be explained in more detail by summarizing the corresponding sections of IEC 61508 (IEC 61508 S+ 2010).

11.8.1   Phase 1: Concept

The goal of the concept phase is to get a sufficient understanding of the EUC and its environment. In this phase, information about the physical environment, hazard sources, hazard situations, and legal requirements should be collected and documented (Thielsch 2012). This includes hazards that occur in the interaction of the system with other systems.

Appropriate methods are, for example, graphics, wiring diagrams, UML/ SysML models, preliminary hazard list, and hazard list.

11.8.2   Phase 2: Overall Scope Definition

The objectives of Phase 2 are to define the boundaries of the system (EUC) and to determine the scope of the hazard and risk analysis. The external, internal, temporal, and probabilistic boundaries of the system should be defined. It should be determined what should be part of the analysis. This refers to the systems itself, other systems, external events, types of failures that should be considered, etc.

Appropriate methods are, for example, consultation with the client/user, use case diagrams, and a finer hazard list.

11.8.3   Phase 3: Hazard and Risk Analysis

In Phase 3, hazard events related to the EUC are determined, sequences of events that lead to those hazard events are identified, and risks that are connected to the hazard are also determined. A hazard and risk analysis based on Phase 2 should be executed and adapted if more information is available later in the development process. The risk should be reduced according to the results of the analysis.

An appropriate system analysis method should be chosen and executed. The system analysis should consider the correctly functioning system, failures, unintentional and intentional incorrect use of the system, etc. Probabilities and consequences of failures, including credible incorrect use should also be included in the analysis.

The sequences of events from Phase 2 should be analyzed in more detail. The risk of the system for different hazards should be determined. Appropriate methods are, for example, hazard analysis, fault tree analysis, FMEA, QRA (quantitative risk analysis), and PRA (probabilistic risk analysis).

11.8.4   Phase 4: Overall Safety Requirements

Figure 3.2 shows the risk concepts of the IEC 61508. By evaluating the risk, it can be determined how safe the EUC is and the overall safety requirements can be derived. The determined requirements have the goal to reduce the EUC risk to a tolerable level by external risk minimization measures and by safety-related systems (Thielsch 2012).

The requirements should be defied in terms of safety functions and safety integrity. "The overall safety functions to be performed will not, at this stage, be specified in technology-specific terms since the method and technology implementation of the overall safety functions will not be known until later" (IEC 61508 2010).

11.8.5   Phase 5: Overall Safety Requirements Allocation

"The first objective of the requirements of this subclause is to allocate the overall safety functions [...] to the designated E/E/PE safety-related systems and other risk reduction measures" (IEC 61508 2010). Secondly, the target failure measure (depending on low demand mode/high demand mode/continuous mode of operation) and the SIL for each safety function are allocated.

11.8.6   Phases 6 to 8: Overall Operation and Maintenance Planning, Overall Safety Validation Planning, and Overall Installation and Commissioning Planning

Phases 6 to 8 will not be explained here. They are neither typical application areas for the methods from the previous chapters nor are they different from the first version of the standard. For more detail on those phases, please refer to the IEC 61508.

11.8.7   Phase 9: E/E/PE System Safety Requirements Specification

Phase 9 from the first version of the IEC 61508 used to be the realization phase where the software and hardware safety life cycle are executed. In Version 2 of the IEC 61508, this is now located in Phase 10. Phase 9 now defines the E/E/PE system safety requirements. They are based on the overall safety requirements defined in Phase 4 and allocated in Phase 5.

The requirements specification should now "provide comprehensive detailed requirements sufficient for the design and development of the E/E/PE related systems" (IEC 61508 2010) and "[e]quipment designers can use the specification as a basis for selecting the equipment and architecture" (IEC 61508 2010).

11.8.8   Phase 10: E/E/PE Safety-Related Systems: Realization

In the first version of IEC 61508, Phase 9 contained the software and hardware safety life cycle. In the second version of the standard, this phase is now Phase 10. Phase 10 of the safety life cycle can again be described in terms of development life cycles, the E/E/PE system hardware development process, and the software development process, for which, respectively, the requirements of V-Models need to be fulfilled.

11.8.8.1   E/E/PE System Life Cycle

The E/E/PE system life cycle is shown in Fig. 11.3.

11.8.8.2   Software Life Cycle

The software life cycle is shown in Fig. 11.4. It interacts with the similarly structured E/E/PE system life cycle from Fig. 11.2.

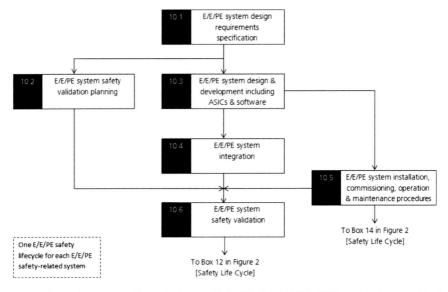

Fig. 11.3 E/E/PE system life cycle from IEC 61508 (IEC 61508 2010) with minor graphical changes. Copyright © 2010 IEC Geneva, Switzerland. www.iec.ch

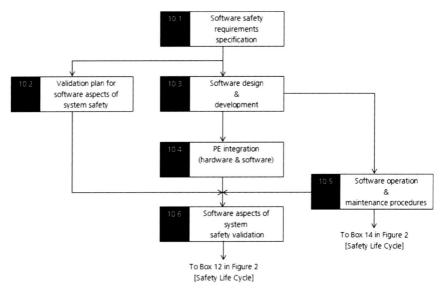

Fig. 11.4 Software life cycle from IEC 61508 (IEC 61508 2010) with minor graphical changes. Copyright © 2010 IEC Geneva, Switzerland. www.iec.ch

Table 11.2 Extract from Table A.1 in IEC 61508 (IEC 61508 2010). Copyright © 2010 IEC Geneva, Switzerland. www.iec.ch

Technique/Measure	Ref.	SIL 1	SIL 2	SIL 3	SIL 4
Semi-formal methods	Table B.7	R	R	HR	HR
Formal methods	B.2.2, C 2.4	–	R	R	HR
Forward traceability between [...]	C.2.11	R	R	HR	HR

For the different phases, there are tables in the Annex of IEC 61508 part 3 indicating which methods should be used depending on the SIL. For each method, the standard distinguishes between highly recommendable (HR), recommendable (R), no recommendation for or against the measure (–), and not recommendable (NR).

The standard provides methods (techniques and measures) for systematic and statistical hardware failures and for systematic software failures. The methods cover prevention of failures as well as control in case of occurrence. Statistical hardware failures are, for instance, opening of solder connections or degradations of components due to vibrations.

Example Table A.1 treats the software safety requirements specification, that is, Phase 10.1. An extract of the table is shown in Table 11.2. These are, in particular, methods for prevention of systematic hardware and software failures.

"Provided that the software safety lifecycle satisfies the requirements of Table 11.1 [listing objectives, scope, requirements subclause, inputs and outputs of the steps in the software life cycle], it is acceptable to tailor the V-model (see Fig. 10.4) to take account of the safety integrity and the complexity of the project" (IEC 61508 2010).

Actual development and safety assessment process steps as well as techniques and measures recommended for software safety can be found to be rather similar when comparing approaches and methods recommended by IEC 61508 with other standards. An example comparison approach and its conduction is provided in (Larisch et al. 2009). It shows that compliance listings are useful for comparison.

11.8.9 Phases 11 to 16: Other Risk Reduction Measures, Overall Installation and Commissioning, Overall Safety Validation, Overall Operation Maintenance and Repair, Overall Modification and Retrofit, and Decommissioning Or Disposal

Phases 11 to 16 will not be explained here. They are neither typical application areas for the methods from the previous chapters nor are they different from the first version of the standard. For more detail on those phases, please refer to the IEC 61508.

11.9   Summary of Requirements for Safety Functions

Finally, some already discussed and missing concepts are listed that need to be fulfilled for each safety functions. A safety function needs to fulfill the following requirements:

- Qualitative descriptions of what the safety function should achieve including time limits and transitions from unsafe to safe state or the maintenance of a safe state that is required.
- Quantification of the reliability of the safety function on demand or continuously, i.e., determination of expected SIL level.
- For the safety architecture of the safety-related system (e.g., serial system or parallel system or in combination) as achieved within the allocation of the safety function for each element, the SIL level must be provided.
- The hardware failure tolerance (HFT) must be sufficient for the level complexity of the elements. One distinguishes between components for which all failure modes are known (level A of complexity) and complex components (level B) for which not all failure modes are known. For instance, $HFT = 1$ means that any single element of the safety-related system may fail without failure of the overall safety-related system.
- The diagnostic coverage (DC) must fit to the level of complexity of the component, the HFT, and the intended SIL.
- For the hardware and software development of each element as well as all the assessment tasks, the appropriate methods must be selected as recommended for the selected SIL level.

11.10   Questions

(1)   Define safety life cycle according to IEC 61508 and name import phases that lead to the specification of the safety function.
(2)   Which quantities need to be provided for each safety function?
(3)   What are techniques and measures (methods) in the context of functional safety?
(4)   How can the quantities SIL, DC/SFF, and HFT be determined?
(5)   Which development model for HW and SW is recommended by IEC 61508?
(6)   Classify into statistical and systematic failures: (a) radiation damage of HW, (b) SW error, and (c) HW component selection does not fit to environmental stress.
(7)   Of how many parts does the standard IEC 61508 consist of?
(8)   When was the second version of the IEC 61508 published?
(9)   How many parts of the standard are normative?
(10) Relate the terms hazardous event, hazardous situation, hazard, and harm to each other.

(11) Give an example of a qualitative and a quantitative description of a safety function (other than the one in the lecture notes).

(12) To which of the safety life cycle phases should the descriptions on the right be related?

 – Understanding of the physical environment,
 – Definition of external boundaries,
 – Determination of EUC-related hazards,
 – Safety requirements based on risk concept,
 – Target failure measure and SIL for each safety function,
 – Safety requirements and technology-specific terms, and
 – Software life cycle.

(13) In which categories does the IEC 61508 categorize the methods? Write down the abbreviations in the order from the highest to the lowest ranking.

11.11 Answers

(1) See Sect. 3.14 and Fig. 11.2 focusing on the early phases and their description in the text.

(2) For each safety function, a qualitative and quantitative description needs to be provided. The qualitative description covers what is done to reduce a risk. It may contain application-specific quantities. Example: "Open airbag within 100 ms after detection of collision." The quantitative description of a safety function is given by the safety integrity level (SIL), defined as the probability of failure on demand or continuously of the safety function.

When realizing a safety function by a related EEPE system, minimum combinations of (i) hardware failure tolerance (HFT), (ii) diagnostic coverage (DC) or safe failure fraction (SFF), and (iii) complexity of components used in EEPE system need to be fulfilled. The Standard IEC 61508 contains best practice tables where the maximum achievable SIL is provided given HFT, Level A/B, and DC. High SILs are only achievable for high HFT, Level A, and high DC.

(3) They are methods that are recommended or should be used to fulfill general or step-specific requirements of the Safety Life Cycle and the Development Processes of the IEC 61508.

(4) The quantities SIL, DC/SFF, and HFT can be determined using FTA and approximated when using FMEDA or DFM. HFT can only be determined with FTA for HFT > 2.

(5) V-Model.

(6) (a) statistical hardware failure, (b) systematic software failure, and (c) systematic hardware design failure.

(7) Seven.

(8) 2010.

(9)   4 (Part 1 to 4).
(10) Hazard = potential source of harm,
      Hazardous situation = circumstance in which a person is exposed to hazard(s),
      and.
      Hazard event = hazardous situation which results in harm.

(11) Qualitative description: Stop of heating of chemical reactor if temperature
      reaches more than 150 degrees. Quantitative description: SIL 3.
(12) The relations are

      – Understanding of the physical environment—Phase 1;
      – Definition of external boundaries—Phase 2;
      – Determination of EUC-related hazards—Phase 3;
      – Safety requirements based on risk concept—Phase 4;
      – Target failure measure and SIL for each safety function—Phase 5;
      – Safety requirements and technology-specific terms—Phase 9; and
      – Software life cycle—Phase 10.

(13) HR,R, –, NR.

References

Bell, R. (2005). Introduction to IEC 61508. Tenth Australian Workshop on Safety-Related
   Programmable Systems.
Börcsök, J. (2006). Funktionale Sicherheit. Heidelberg, Hüthig Verlag.
Hänle, A. (2007). Modellierung und Spezifikation von Anforderungen eines sicherheitskritischen
   Systems mit UML, Modeling and Specification of Requirements of a safety critical System with
   UML. Diploma Thesis, Hochschule Konstanz für Technik, Wirtschaft und Gestaltung (HTWG),
   University of Applied Sciences; Fraunhofer EMI, Efringen-Kirchen.
IEC 61508 (2010). Functional Safety of Electrical/Electronic/Programmable Electronic Safety-
   related Systems Edition 2.0 Geneva, International Electrotechnical Commission.
IEC 61508 S+ (2010). Functional Safety of Electrical/Electronic/Programmable Electronic Safety-
   related Systems Ed. 2 Geneva, International Electrotechnical Commission.
Larisch, M., A. Hänle, I. Häring and U. Siebold (2008a). Unterstützung des Nachweises funktionaler
   Sicherheit nach IEC 61508 durch SysML. Dipl. Inform. (FH), HTWG-Konstanz.
Larisch, M., A. Hänle, U. Siebold and I. Häring (2008b). SysML aided functional safety assessment.
   Safety Reliablity and Risk Analysis: Theory, Methods and Applications, European Safety and
   Reliablity Conference (ESREL) 2008. S. Martorell, C. G. Soares and J. Barett. Valencia, Spanien,
   Taylor and Franzis Group, London. 2: 1547–1554.
Larisch, Mathias; Siebold, Uli; Häring, Ivo (2009): Principles of the AOP 52 draft on software
   safety for the ammunition domain. In: European Safety and Reliablity Conference (ESREL)
   2009. Prague, Czech Republic.: Taylor and Franzis Group, London, pp. 1347–1352.
Liggesmeyer, P. and D. Rombach (2005). Software Engineering eingebetteter Systeme. München,
   Elsevier Spektrum Akademischer Verlag.
Redmill, F. J. (1998a). "IEC 61508 - Principles and use in the management of safety." Computer
   and Control Engineering Journal 9(5): 205–213.
Redmill, F. J. (1998b). "An Introduction to the Safety Standard IEC 61508." Journal of System
   Safety 35, no. 1(1): 10–22.

Redmill, F. J. (2000a). Installing IEC 61508 and Supporting Its Users - Nine Necessities. Workshop for Safety Critical Systems and Software, Australia.

Redmill, F. J. (2000b). Understanding the Use, Missuse and Abuse of Safety Integrity Levels, *Proceedings of the Eighth Safety-critical Systems Symposium*, pp. 20–34, Springer.

Schmidt, Andreas; Häring, Ivo (2007): Ex-post assessment of the software quality of an embedded system. In Terje Aven, Jan Erik Vinnem (Eds.): Risk, Reliablity and Societal Safety, European Safety and Reliability Conference (ESREL) 2007, vol. 2. Stavangar, Norway: Taylor and Francis Group, London, pp. 1739–1746.

Smith, D. J. and K. G. Simpson (2004). Functional Safety - A Straightforward Guide to Applying IEC 61508 and Related Standards. London, Elsevier.

Thielsch, P. (2012). Risikoanalysemethoden zur Festlegung der Gesamtssicherheitsanforderungen im Sinn der "IEC 61508 (Ed. 2)". Bachelor, Hochschule Furtwangen.

Chapter 12
Requirements for Safety-Critical Systems

12.1 Overview

The identification of existing safety functions of established and legacy systems, the determination of requirements for standard safety functions, and especially the development of innovative and resource-efficient safety functions are key for the development of efficient and sustainable safety-relevant and safety-critical systems.

For instance, it is not yet clear which functions of autonomous driving can be considered as reliability functions (intended functions) and which need to be considered safety-critical functions, or both. In this case, it is obvious that tremendous economic, societal, and individual interests drive such safety-critical system developments and introductions. An example of an attempt to standardize parts of the verification and validation of automotive intended system functions is given by the Safety of the Intended Functionality standard (ISO 21448 2019), which is complementary to (ISO 26262 Parts 1 to 12 Ed. 2 2018), itself an application standard of IEC 61508 S+ (2010).

Safety-related functions even in standard situations need approximately doubled resources when compared to standard system developments. Therefore, it is important to identify sufficient, resource-efficient, economically, societally, and legally accepted safety functions. This includes to take advantage of any possible innovations to develop them.

To this end, this chapter treats properties (dimensions, aspects) of safety requirements. Several, mostly pairs of, adjectives are listed with which safety requirements can be classified. It is also addressed which combinations are likely to appear. With the help of the properties and corresponding adjectives the search for possible safety functions in a given context can be structured. In addition, one can search for possible safety functions that have not yet been considered. Such safety functions might be very successful but not yet used due to technological gaps.

Next the chapter lists examples for safety requirements and the attributes are assigned, i.e., which attributes fit to the safety requirements. It is concluded with combinations of properties are likely to appear and which are not yet often used.

© The Author(s), under exclusive license to Springer Nature Singapore Pte Ltd. 2021
I. Häring, *Technical Safety, Reliability and Resilience*,
https://doi.org/10.1007/978-981-33-4272-9_12

In detail, the chapter covers the following contents. Section 12.2 further explains the importance of safety requirements and their role in system development processes.

Section 12.3 gives a list of definitions which are important for this chapter. The definitions of risk and safety are repeated. Harm, hazardous events and situations, hazard, safety functions, highly available systems and safety-critical system, and safety requirements are introduced in parts with slight variations when compared to the definitions already given in Chaps. 3 and 9, due to the fact that safety requirements beyond the context of functional safety are included in the present chapter.

Section 12.4 lists in 10 subsections properties of safety requirements. Most of them are defined in opposing pairs. Examples are given. Section 12.4.11 does not directly give properties of safety requirements but system safety properties such as reliability or availability which can play an important role for safety requirements.

Section 12.5 determines which combinations of properties of Sect. 12.4 appear most often, based on the examples of Sect. 12.4. Typical combinations are, for instance, safety requirements of structural nature (non-functional) that aim at preventing damage events which are specified in detail. Another combination often encountered is functional safety requirements that aim at mitigating damage effects during or post event occurrence which may be specifically focused on the hazard cause or effect. In contrary, functional non-specific safety requirements before events do not occur very often.

In summary, the aim of this chapter is to introduce terms for a clear and comprehensible characterization (classification) of safety requirements and to compare which combination of terms typically appear in combination with each other. This leads to the classification of safety requirements by their properties as well as potential promising combinations of such properties which are not employed very often.

The chapter is a summary, translation, and partial paraphrasing and extension of (Hänle 2007; Hänle and Häring 2008).

12.2   Context

Many accidents occur due to problems with system requirements for hardware and software. In particular, empirically based statements confirm the generally assumed hypothesis that the majority of safety problems for software originates in software requirements and not in programming errors (Firesmith 2004).

The requirements are often incomplete (Hull et al. 2004). They normally do not specify prevention of rare dangerous events. The specification often also does not describe how one should react in exceptional situations. In the cases where dangers and exceptions are described, the requirements usually are not specific enough or parts are missing which is only noticed by specialists in that work field (Firesmith 2004).

Earlier safety technologies were invented as a reaction to technological innovations. Nowadays, it is often also the other way around. Technological innovations take place due to safety requirements. Practice has shown that it is more expensive to install safety later on than to include it as part of the concept from the start (Zypries 2003).

12.3 Definitions

12.3.1 Safety and Risk

Risk and safety have already been defined in Sects. 3.4 and 3.7. As a reminder, safety is defined as "freedom from unacceptable risks" (IEC 61508 2010) and following the classical definition of risk by Blaise Pascal we define the risk as the product of probability and consequences: $R = PC$.

12.3.2 Highly Available and Safety-Critical Systems

Highly available systems are systems that have to work reliably. A failure of the system can cause high damage. Safety-critical systems can also be highly available (Hänle 2007).

For *safety-critical* systems, there is always the risk of a hazardous event involving persons. A highly available system may be free of risk for persons and "only" lead to financial damage. For example, the failure of a bank's data center can lead to costs of millions per hour (Lenz 2007), but does not cause harm on persons.

12.3.3 Safety Requirement

We define a *safety requirement* as "a requirement for a system whose operation causes a non-acceptable risk for humans and the environment. The fulfillment of the safety requirement reduces the risk associated with the system" (Hänle and Häring 2008).

12.4 Properties of Safety Requirements

This section introduces properties of safety requirements as explained in Hänle (2007) and Hänle and Häring (2008). "A very interesting work and similar approach in this area was written by Glinz (2005, 2007). He calls his approach faceted classification" (Hänle and Häring 2008).

Further examples of literature on properties of safety requirements are Lano and Bicarregui (1998), Smarandache and Nissanke (1999), Tsumaki and Morisawa (2001), Firesmith (2004), Grunske (2004), Ober et al. (2005), Fontan et al. (2006), Freitas et al. (2007) and for the property "functionality" also Antón (1997), Jacobson et al. (1999), Robertson and Robertson (1999), Glinz (2006, 2007), Sommerville (2007).

12.4.1 Functional Versus Non-functional Safety Requirements

IEC 61508 distinguishes between safety requirements for functional safety and non-functional safety. *Functional safety* includes every function of the system which is executed to guarantee the safety of the system. *Non-functional safety* is fulfilled by properties, parts, or concepts of the system which guarantee safety by their existence (Hänle 2007).

12.4.2 Active Versus Passive Safety Requirements

The automotive sector uses the terms active and passive safety (Exxon Mobil Corporation 2007; Hänle 2007). *Active safety* is anything that prevents accidents. This comprises effective brakes, smooth steering, antilock brakes, fatigue-proof seats, and clearly arranged and uncomplicated control and indicating elements.

Passive safety requirements are requirements for those functions and properties of the system which prevent or minimize damage in case of an accident. Examples for those functions/properties are airbags, emergency tensioning seat belt retractors, and deformation zones (Hänle 2007).

This shows that the classification into active and passive is not the same as the one into functional and non-functional although this might first be expected by the choice of terms. The difference can also be seen in the following examples.

Examples (Hänle 2007):

– Active functional safety requirements are safety requirements for the functions of the system that have to be executed as positive action to prevent a dangerous situation, for example, turning off the engine (DIN IEC 61508 2005).
– Passive functional safety requirements are safety requirements for safety functions that have a surveillance function and only interfere if there is a problem. An example is the requirement to prevent a start of the engine in case of a dangerous event (DIN IEC 61508 2005).
– An example for an active non-functional safety requirement is to provide a special insulation to withstand high temperatures (DIN IEC 61508 2005).

- Another example is the use of redundant components which still guarantee the correct functioning of the system in case of a failure in one component. This can be an example for an active non-functional as well as for a passive non-functional safety requirement, depending on where the concept is used.

12.4.3 Technical Versus Non-technical Safety Requirements

Technical safety requirements state how something has to be technically realized. Examples are Hänle (2007)

- The components of a safety system should be located in a single device, generalized from (STANAG 4187 Ed. 3 2001).
- Systems of category III should contain redundant equipment in order to still function in case of a single failure (Civil Aviation Authority 2003).
- Sensors and safety devices should be located in different places.

Requirements which do not state how something should be technically realized in the system are *non-technical* requirements. Examples are (Hänle 2007)

- The system should fulfill abstract design principles of a standard, e.g., (STANAG 4187 Ed. 3 2001).
- The airbag should have inflated after a specified time (Benoc 2004).

12.4.4 Concrete Versus Abstract Safety Requirements

Abstract safety requirements describe general properties. *Concrete* safety requirements precisely describe how something has to be realized. Those can be technical but also non-technical, for example, physical or temporal requirements (Hänle 2007).

In comparison to technical requirements, concrete safety requirements do not have to say anything about the constructive realization.

Example (Hänle 2007): The requirement that an airbag should inflate after a specified time is an abstract, general definition. A concrete (but non-technical) requirement would be that the airbag should inflate 30–40 ms after the collision was detected (Hänle 2007).

Distinguishing between abstract and concrete safety requirements is not always easy to define because the transition from one to the other is smooth. A requirement in between the two from the example would be that the airbag should inflate during a specified time period after the collision was detected consistent with the dynamic passenger movement that should be controlled, however, without specifying how long this time period exactly is Hänle (2007).

12.4.5 Cause- Versus Effect-Oriented Safety Requirements

For safety-critical systems, risks determine the major part of the safety requirements (Storey 1996). A "Safety Requirement" document is usually created to describe how the system has to be secured to guarantee safety (Hänle 2007).

Risk analysis of a system identifies risks and leads to the definition of safety requirements for the system (Grunske 2004). According to Leveson (1995), the requirements defined in order to prevent identified risks together with the quantitative determination of tolerable hazard rates for each risk form the safety requirements for the system (Hänle 2007).

Here two types of safety requirements can be distinguished. The one type refers to which risks have to be regarded and the other to how the risks can be prevented. We will call the first a *cause-oriented* safety requirement and the latter an *effect-oriented* safety requirement (Hänle 2007).

Effect-oriented safety requirements can partly consist of cause-oriented safety requirements. The cause-oriented requirements are used to name the reason for the effect-oriented requirement. Those composed safety requirements are counted as effect-oriented because effect-oriented requirements are stronger requirements for the realization than cause-oriented requirements (Hänle 2007).

Example (Hänle 2007): Consider a storage room. A cause-oriented safety requirement is that a person should not be locked in the room. An effect-oriented safety requirement is that the room should contain an infrared sensor which prevents a closing of the door when it detects a person in the room. An effect-oriented safety requirement with a cause-oriented part is that the room should contain an infrared sensor which prevents a closing of the door when it detects a person in the room to prevent locking a person in.

Example (Hänle 2007): A cause-oriented safety requirement is that pressure equipment has to be designed, tested, and, if necessary, equipped and installed in such a way that safety is guaranteed if it is used in accordance to the producer's requirements and under reasonably predictable conditions (Richtlinie 97/23/EG 1997).

Example (Hänle 2007): An effect-oriented safety requirement with a cause-oriented part is: Where necessary, adequate means to drain and vent pressure equipment have to be provided

– to prevent damaging effects like water hammer, vacuum failures, corrosion, and uncontrolled chemical reactions and
– for safe cleaning, inspection, and maintenance (Richtlinie 97/23/EG 1997).

12.4.6 Static Versus Dynamic Safety Requirements

Static safety requirements are requirements which have to be fulfilled during the product's entire life cycle, for example, constructive safety requirements or quality requirements (Hänle 2007).

It is rather easy to verify during or after the development process that static safety requirements are fulfilled. For this, it is essential to know the whole system (Hänle 2007).

Examples (Hänle 2007):
Examples for static safety requirements are

- Quality requirement: The product should be developed according to Norm XY. This can easily be proven: Either Norm XY was used during the developing process or it was not.
- The emergency shutoff should always have priority (DIN EN 418 1993).

Dynamic safety requirements are requirements which do not have to be fulfilled during the entire life cycle. This includes requirements that have to be fulfilled regularly, several times, or once during the product's time of use. Hence, their fulfillment can only be proven in certain time spans.

In contrast to static safety requirements, the fulfillment of dynamic requirements cannot be proven as easily and often not for every situation.

Examples (Hänle 2007):
Examples for dynamic safety requirements are

- The reaction of a machine to the emergency shutoff is not allowed to cause further hazard (DIN EN 418 1993).
- The airbag should open in such a way that persons are only little at risk.

It often is not possible to prove that a requirement is fulfilled correctly in all dynamic situations due to the complexity of modern systems. So, only a certain number of possible system runs can be tested and a fault can still be undetected after the test (Buschermöhle et al. 2004).

A formal description of the requirements can be used for a formal proof of correctness for all possible system runs. This can be very elaborate and expensive (Buschermöhle et al. 2004; Hänle 2007).

12.4.7 Standardized Requirements

Under the term *standardized* we summarize all safety requirements where the fulfillment yields additional safety or reliability. Typically, those requirements ask for approved production processes and methods (Hänle 2007).

Examples (Hänle 2007):
Examples for standardized safety requirements are

– The usage of adequate norms.
– The usage of approved components or construction rules, for example, "Reactivity control devices should be used to maintain the reactor in a subcritical condition, with account taken of possible design basis accidents and their consequences" (Safety Guide NS-G-1.12 2005).
– The usage of approved safety principles, for example, stopping the system in case of failure (Grell 2003).

Standardized requirements are only a small part of the safety requirements. Many requirements are not standardized, for example, the functional safety requirements of IEC 61508.

12.4.8 Qualitative Versus Quantitative Safety Requirements

Quantitative safety requirements describe requirements with one performance feature that a system or a part of a system at least has to fulfill. According to Glinz (2006), quantitative requirements can be distinguished between requirements for

– time,
– quantity,
– capacity,
– precision, and
– rates (e.g., transaction rate, transfer rate).

Examples (Hänle 2007):
Examples for quantitative safety requirements are

– Permissible loading of certain pressure equipment: 1% elongation limit for unalloyed aluminum, translated from Richtlinie 97/23/EG (1997).
– Reliability: Avionic systems are not allowed to have more than one catastrophic failure in a billion hours (frequency $10^{-9}h^{-1}$) (Grell 2003).
– Precision: "The radar recording time source shall be synchronized with the main station time source to a tolerance of within ± 5 s."
– Time: "In designing for the rate of shutdown, the response time of the protection systems and the associated safety actuation systems (the means of shutdown) should be taken into account" (Civil Aviation Authority 2003; Safety Guide NS-G-1.12 2005).

Qualitative safety requirements describe the way in which a system fulfills safety to minimize or prevent the hazard of persons and environment in a verbose way, for example, that a machine stops if the protection of the blade is removed (Hänle 2007).

12.4.9 System-Specific Versus Module-Specific Safety Requirements

System-specific safety requirements are requirements which refer to the whole system (Ober et al. 2005).

Module-specific safety requirements only refer to parts of the system, down to single elements.

12.4.10 Time-Critical Safety Requirements

There are safety requirements which ask for something to occur in a certain period of time. The velocity with which the task is completed is not very important as long as the task is finished in the fixed time period (Stallings 2002). We call those safety requirements *time-critical*.

Examples (Hänle 2007):
Examples for time-critical safety requirements are

– The blade has to stop after at most 1 s when the protection is removed (DIN IEC 61508 2005).
– The railroad-crossing gate must be closed before the train reaches a safety distance of 250 meters (Beyer 2002).

An example for a time-uncritical safety requirement would be to require a certain construction method (Hänle 2007).

12.4.11 System Safety Properties

According to Storey (1996), there are seven different properties of system safety. Their importance varies with the specific system under consideration and with this their influence on the safety requirements for the system.

Those seven properties are reliability, availability, fail-safe state, system integrity, data integrity, system recovery, and maintainability (Hänle 2007).

Reliability means that a function should work correctly over a certain period of time (minutes, hours, days, years, decades). An example for such a system is a machine used in medicine to monitor the cardiac function (Hänle 2007). See the example given in Sect. 4.12.

Availability requires that a system works when it is needed. An example is the emergency shutdown of a nuclear power plant. It is not required that this function works correctly for a long period of time but that it works sufficiently close to the point in time when it is needed (Hänle 2007). See also Sect. 9.2.

Some systems can be in states which are regarded as safe. *Fail-safe* means that the system transfers to such a safe state in case of a failure (Ehrenberger 2002). An example is the stopping of a train in case of a failure. Some systems do not have safe states, for example, an airplane while it is in the air (Hänle 2007).

System integrity describes the property of a system to detect and display failures in its own operations. An example is an autopilot system which shows a warning if it doubts the correct execution of its own functions (Hänle 2007).

Data integrity describes the property of a system to prevent, recognize, and possibly correct mistakes in its own database (including file-IO). Incorrect data can lead to wrong decisions of the system. An example for data integrity is to store data three times and only use it if two stored values are exactly the same (Hänle 2007).

Especially for systems without safe state it is important to be able to restore the system (*system recovery*). It depends on the type of system whether this is possible. An example would be to restart a system after a failure (Hänle 2007).

Maintainability also has to be considered because not all systems can be maintained regularly or even at all. This should be regarded when deciding on safety requirements. An example for a system that cannot be maintained is a satellite (Hänle 2007).

Examples (Hänle 2007):

- Reliability and availability: "Duplicate data paths should be implemented to increase the availability" (Civil Aviation Authority 2003).
- Availability: The emergency shutoff should always be available and executable, independent of the operation mode (DIN EN 418 1993).
- Data integrity: "The device used for radar recording shall not be capable of erasing any recorded data" (Civil Aviation Authority 2003).
- Maintainability: Emergency power systems (EPSs) for nuclear power plants "that supply electrical and non-electrical power to systems important to safety are of fundamental importance to the safety of nuclear power plants. The purpose of the EPSs is to provide the plant with the necessary power in all relevant conditions within the design basis so that the plant can be maintained in a safe state after postulated initiating events, in particular during the loss of off-site (grid) power" (Safety Guide NS-G-1.8 2005).

12.5 Evaluating the Properties

The properties of safety requirements are summarized in Table 12.1.

Hänle (2007), Hänle and Häring (2008) used the examples from Sect. 12.4 and some additional ones (32 examples in total) to evaluate how often certain combinations of the properties from Sect. 12.4.11 occur in those safety requirements. He found his examples "by analyzing various safety requirements (European Directive 70/311/EWG 1970; DIN EN 418 1993; European Directive 97/23/EG 1997; IEC 61508 1998-2005 (2005); STANAG 4187 Ed. 3 2001; Beyer 2002; Civil Aviation

Table 12.1 Properties of safety requirements. If two properties are not separated by a line only one property may be simultaneously assigned to a safety requirement (Hänle and Häring 2008)

Property	A safety requirement has this property
Functional	If the system should have a specific function or behavior
Non-functional	If the system should have a specific characteristic or a specific concept should be used
Active	If something is required that prevents a hazardous event
Passive	If something is required that mitigates the consequences of an occurring hazardous event
Concrete/precise	If the requirement has detailed information
Abstract	If the requirement is generally described
Technical	If the requirement describes how something has to be realized
Non-technical	If the requirement does not describe how something should be realized
Cause-oriented	If it describes one or more hazards that the system should prevent
Effect-oriented	If something is described that mitigates or prevents hazards
Static	If the requirement may be verified at any time
Dynamic	If the requirement may only be verified at a specific point in time in the product life cycle of the system
Quantitative	If a numerical or a proportional specification is given
Qualitative	If a verbal requirement describes the method how a system satisfies safety
Standardized	If the use of something proven is requested, e.g., the use of a standard, proven components, proven procedures, prohibition of something specific
System-specific	If the requirement applies to a system as a whole
Module-/component-specific	If the requirement applies only to a part of the system
Time-critical	If a task must be fulfilled till a specific date or within a specific time period
Time-uncritical	If it is not time-critical
Reliability	If the requirement demands the reliable functioning on demand
Availability	If the requirement calls for the functioning of a system function in the correct moment
Fail-safe state	If the requirement calls for a safe state in which the system can be set in case of a failure. This is also called »fail-safe« (Ehrenberger 2002)
System integrity	If the requirement demands self-analyzing capability so that the system can recognize failures of its own operations
Data integrity	If the requirement demands recognition of failures in the database of the system
System recovery	If the requirement calls for a reset function so that the (sub-)system can be recovered from a failure state
Maintainability	If the requirement describes how the system can be maintained

Authority 2003; Grell 2003; Benoc 2004; DIN EN ISO 12100 2004; Safety Guide NS-G-1.8 2005; Safety Guide NS-G-1.12 2005) and literature (Leveson 1995; Storey 1996; Lano and Bicarregui 1998; Jacobson et al. 1999; Robertson and Robertson 1999; Lamsweerde 2000; Tsumaki and Morisawa 2001; Bitsch and Göhner 2002; Firesmith 2004; Grunske 2004; Glinz 2005, 2007; Fontan et al. 2006; Ober et al. 2006, Freitas et al. 2007; Sommerville 2007)" (Hänle and Häring 2008).

The analysis was executed in a spreadsheet application (Microsoft Excel) using "AutoFilter" based on a table as in the representative Table 12.2 with one line for each of the 32 examples. In the table, "−" means that the property is not relevant and "?" that it is not clear to which of the properties it should be related.

Due to the small dataset of 32 safety requirements, the evaluation cannot be seen as a proof of which combinations usually occur but the results can be seen as an indicator. Hänle's suggestions are

− Functional passive and non-functional active safety requirements occur much more often than functional active and non-functional passive requirements, see Fig. 12.1. This means that more safety requirements are defined to require functions which intervene in case of an accident in order to minimize the damage rather than to approach accidents in a functional active way. On the other hand, properties that are defined for the system usually prevent accidents, such as special insulation or high-quality components. Properties that reduce damage in case of an accident, for example, deformation zones in cars, are rare.
− Non-functional effect-oriented safety requirements are static which implies that they can always be tested and they are always time-uncritical.
− Cause-oriented safety requirements are, due to their description of risks, non-functional abstract non-technical dynamic safety requirements.
− Time-critical safety requirements always are functional dynamic non-technical safety requirements. This is supported by Liggesmeyer (2002) who classifies time-critical requirements as functional properties of a technical system.
− Technical safety requirements are concrete time-uncritical requirements. Non-technical requirements can be concrete or abstract.
− Dynamic safety requirements, that is, requirements that can only be tested at certain time points, normally are functional safety requirements but not always.
− Standardized safety requirements are static safety requirements. This means that they can be verified at any time point in the system life cycle (possibly using development documentation documents).
− Concrete/abstract and system-/module-specific properties do not show accumulations and can be associated with any type of requirement.
− Due to the small dataset and the long list of system safety properties, one cannot say much about the association of safety requirements to those properties. It can be only be seen that safety requirements with the system safety property "availability" normally are functional and safety requirements with the system safety property "reliability" normally are non-functional.

Table 12.2 Evaluation of the properties of the example safety properties (Hänle 2007)

Safety requirement	Functional (f)/non-functional (nf)	Active (a)/passive (p)	Concrete (c)/abstract (a)	Technical (t)/non-technical (nt)	Cause- (c)/effect-oriented (e)	Static (s)/dynamic (d)	Qualitative (ql)/quantitative (qt)	Standardized (s)	System (s)-/module-specific (m)	Time-critical (c)/time-uncritical (u)	System safety property
Category III systems (see Sect. 12.4.3)	nf	p	c	t	e	s	ql	s	S	u	rel[a]
Single device (see Sect. 12.4.3)	nf	a	c	t	e	s	ql	s	M	u	rel[a]
Airbag in 30–40 ms (see Sect. 12.4.3)	F	p	c	nt	e	d	qt	–	M	c	av[a]
Airbag without time restriction (see Sect. 12.4.6)	F	p	a	nt	e	d	qt	–	M	c	av[a]
Nuclear power plant safety (see Sect. 12.4.11)	nf	?	a	nt	c	d	ql	–	S	u	?

[a]rel = reliability, av = availability

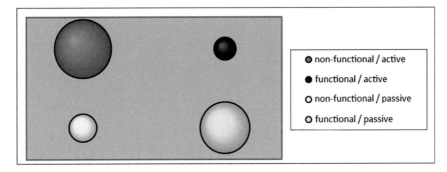

Fig. 12.1 Evaluation of the example safety requirements (Hänle 2007)

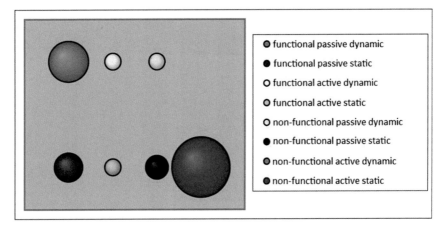

Fig. 12.2 Evaluation of the example safety requirements (Hänle 2007)

– The most common combinations in the analysis of the combination functional/non-functional, active/passive, and static/dynamic where non-functional active static and functional passive dynamic, see Fig. 12.2.

12.6   Questions

(1) Explain the difference between functional and active as presented in this chapter.
(2) Regard the pairs of properties of safety requirements:

   • Functional versus non-functional,
   • Active versus passive,
   • Technical versus non-technical,
   • Concrete versus abstract,
   • Cause-oriented versus effect-oriented,

- Static versus dynamic,
- Qualitative versus quantitative,
- System-specific versus module-specific,
- Time-critical versus time-uncritical.

Define the pairs in each case. Give an example.

(3) For each pair, is it easier to verify a requirement that fulfills the first or the second property of question (2), based on the examples from this chapter? For which properties is it not possible to say? Does this coincide with the experiences from your work?

(4) Classify the following safety requirements by their properties as in Table 12.2:

   - Closing the gate if a train is closer than 300 m.
   - A person should not be locked in a storage room.

(5) The data center of a big company with sensible data about their products is a ... system.

   (a) highly available and safety-critical,
   (b) not highly available and safety-critical,
   (c) highly available and not safety-critical, and
   (d) not highly available and not safety critical.

(6) Name a requirement for a car that is not a safety requirement.

(7) Argue how innovation in terms of novel types of safety functions could look like.

12.7 Answers

(1) Functional refers to functions. They have to be actively executed. Active refers to the prevention of accidents, i.e., active prevention.

(2) See definitions of Sect. 12.4.

(3) Non-functional, active, technical, concrete, static, and time-uncritical should be easier than their opponent. For cause-/effect-oriented, qualitative/quantitative, and system-/module-specific it is hard to say.
   One can also prevent an accident by a safe property of a system (active, non-functional) and an actively executed function can react in case of an accident instead of preventing it (functional, passive).

(4) Closing the gate: functional, passive, concrete, non-technical, effect-oriented, dynamic, quantitative, likely standardized, system-specific, time-critical, and availability.
   Locked person: non-functional, active, abstract, non-technical, cause-oriented, static, qualitative, not standardized, system-specific, time-critical, availability.

(5) (c) A failure in the data center causes high damage but no harm on persons.

(6) Sample functional requirements: Persons to be transported, fast acceleration, heating and cooling system, environmental requirements, and license plate. Please note that steering, sufficient acceleration, and deceleration are (implicitly) parts of safety functions in case of emergencies, e.g., sudden accident on highway.

(7) See the more seldom combinations in Figs. 12.1 and 12.2.

References

Antón, A. (1997). Goal Identification and Refinement in the Specification of Information Systems. PhD.

Benoc. (2004). "Sicherheitsanforderungen an das Airbag System." Retrieved 2007-09-15, from http://www.benoac.de/kompetenzen_sicherheit_de.html.

Beyer, D. (2002). Formale Verifikation von Realzeit-Systemen mittels Cottbus Timed Automata. Berlin, Mensch & Buch Verlag.

Bitsch, F. and P. Göhner (2002). Spezifikation von Sicherheitsanforderungen mit Safety-Patterns. Software Engineering in der industriellen Praxis. Düsseldorf, VDI Verlag GmbH.

Buschermöhle, R., M. Brörkens, I. Brückner, W. Damm, W. Hasselbring, B. Josko, C. Schulte and T. Wolf (2004). "Model Checking - Grundlagen und Praxiserfahrungen." Informatik-Spektrum 27(2): 146–158.

Civil Aviation Authority (2003). CAP 670 - Air Traffic Services Safety Requirements. London, Civil Aviation Authority.

DIN EN 418 (1993). NOT-AUS-Einrichtung, funktionelle Aspekte, Gestaltungsleitsätze. D. E.-S. v. Maschinen. Berlin, Beuth Verlag GmbH.

DIN EN ISO 12100 (2004). Teil 2: Technische Leitsätze. DIN EN ISO 12100 - Sicherheit von Maschinen - Grundbegriffe, allgemeine Gestaltungsleitsätze. Berlin, Beuth Verlag GmbH.

DIN IEC 61508 (2005). Funktionale Sicherheit sicherheitsbezogener elektrischer/elektronischer/programmierbarer elektronischer Systeme, Teile 0 - 7. DIN. Berlin, Beuth Verlag.

Ehrenberger, W. (2002). Software-Verifikation - Verfahren für den Zuverlässigkeitsnachweis von Software. Munich, Vienna, Carl Hanser Verlag.

European Directive 70/311/EWG (1970). Angleichung der Rechtsvorschriften der Mitgliedstaaten über die Lenkanlagen von Kraftfahrzeugen und Kraftfahrzeuganhängern. Richtlinie 70/311/EWG Luxemburg, Der Rat der Europäischen Gemeinschaften.

European Directive 97/23/EG (1997). Anhang I - Grundlegende Sicherheitsanforderungen, Das Europäische Paralament und der Rat der Europäischen Union.

Exxon Mobil Corporation. (2007). "Active Safety Technologies." Retrieved 2012-06-04, from http://www.mobil1.com/USAEnglish/MotorOil/Car_Care/Notes_From_The_Road/Safety_System_Definitons.aspx.

Firesmith, D. G. (2004). "Engineering Safety Requirements, Safety Constraints, and Safety-Critical Requirements." Journal of Object Technology 3(3): 27–42.

Fontan, B., L. Apvrille, P. d. Saqui-Sannes and J. P. Courtiat (2006). Real-Time and Embedded System Verification Based on Formal Requirements. IEEE Symposium on Industrial Embedded Systems (IES'06). Antibes, France.

Freitas, E. P., M. A. Wehrmeister, C. E. Pereira, F. R. Wagner, E. T. S. Jr. and F. C. Carvalho (2007). Using Aspect-oriented Concepts in the Requirements Analysis of Distributed Real-Time Embedded Systems. IFIP International Federation for Information Processing. Boston, Springer.

Glinz, M. (2005). Rethinking the Notion of Non-Functional Requirements. Proceedings of the Third World Congress for Software Quality. Munich, Germany, Department of Informatics, University of Zurich.

Glinz, M. (2006). Requirements Engineering I. Lecture Notes. Universität Zürich.

Glinz, M. (2007). On Non-Functional Requirements. Proceedings of the 15th IEEE International Requirements Engineering Conference. Delhi, India, Department of Informatics, University of Zurich.

Grell, D. (2003). Rad am Draht - Innovationslawine in der Autotechnik. c't - magazin für computertechnik. **14**.

Grunske, L. (2004). Strukturorientierte Optimierung der Qualitätseigenschaften von softwareintensiven technischen Systemen im Architekturentwurf. Potsdam, Hasso Plattner Institute for Software Systems Engineering, University Potsdam.

Hänle, A. (2007). Modellierung und Spezifikation von Anforderungen eines sicherheitskritischen Systems mit UML, Modeling and Specification of Requirements of a safety critical System with UML. Diploma Thesis, Hochschule Konstanz für Technik, Wirtschaft und Gestaltung (HTWG), University of Applied Sciences; Fraunhofer EMI, Efringen-Kirchen.

Hänle, A. and I. Häring (2008). UML safety requirement specification and verification. Safety Reliablity and Risk Analysis: Theory, Methods and Applications, European Safety and Reliablity Conference (ESREL) 2008. S. Martorell, C. G. Soares and J. Barett. Valencia, Spain, Taylor and Franzis Group, London. **2:** 1555–1563.

Hull, E., K. Jackson and J. Dick (2004). Requirements Engineering, Springer.

IEC 61508 (1998-2005). Functional Safety of Electrical/Electronic/Programmable Electronic Safety-related Systems. Geneva, International Electrotechnical Commission.

IEC 61508 (2010). Functional Safety of Electrical/Electronic/Programmable Electronic Safety-related Systems Edition 2.0 Geneva, International Electrotechnical Commission.

IEC 61508 S+ (2010). Functional Safety of Electrical/Electronic/Programmable Electronic Safety-related Systems Ed. 2 Geneva, International Electrotechnical Commission.

ISO 26262 Parts 1 to 12 Ed. 2, 2018-12: Road vehicles - Functional safety. Available online at https://www.iso.org/standard/68383.html.

ISO 21448, 2019-01: Road vehicles — Safety of the intended functionality. Available online at https://www.iso.org/standard/70939.html, checked on 9/21/2020.

Jacobson, I., G. Booch and J. Rumbaugh (1999). Unified Software Development Process. Amsterdam, Addison-Wesley Longman.

Lamsweerde, A. v. (2000). Requirements Engineering in the Year 00: A Research Perspective. ICSE'2000 - 22nd International Conference on Software Engineering. Limerick, Irland, ACM Press.

Lano, K. and J. Bicarregui (1998). Formalising the UML in Structured Temporal Theories. Second ECOOP Workshop on Precise Behavioral Semantics. Brussels, Belgium.

Lenz, U. (2007). "IT-Systeme: Ausfallsicherheit im Kostenvergleich." Retrieved 2012-05-31, from http://www.centralit.de/html/it_management/itinfrastruktur/458076/index1.html.

Leveson, N. G. (1995). Safeware: System Safety and Computers. Boston, Addison-Wesley.

Liggesmeyer, P. (2002). Software-Qualität: Testen, Analysieren und Verifizieren von Software. Heidelberg, Berlin, Spektrum Akademischer Verlag.

Ober, I., S. Graf and I. Ober (2006). "Validating timed UML models by simulation and verification." International Journal on Software Tools for Technology Transfer (STTT) **8**(2): 128.

Ober, I., S. Ober and I. Graf (2005). "Validating timed UML models by simulation and verification." International Journal on Software Tools for Technology Transfer (STTT).

Richtlinie 97/23/EG (1997). Anhang I - Grundlegende Sicherheitsanforderungen, Das Europäische Parlament und der Rat der Europäischen Union.

Robertson, S. and J. Robertson (1999). Mastering the Requirements Process. Amsterdam, Addison-Wesley Longman.

Safety Guide NS-G-1.8 (2005). Design of Emergency Power Systems for Nuclear Power Plants. Safety Guide NS-G-1.8 - IAEA Safety Standards Series. Vienna, International Atomic Energy Agency.

Safety Guide NS-G-1.12 (2005). Design of the Reactor Core for Nuclear Power Plants. Safety Guide NS-G-1.12 - IAEA Safety Standards for protecting people and the environment. Vienna, International Atomic Energy Agency.

Smarandache, I. M. and N. Nissanke (1999). "Applicability of SIGNAL in safety critical system development." IEEE Proceedings - Software **146**(2).

Sommerville, I. (2007). Software Engineering 8. Harlow, Pearson Education Limited.

Stallings, W. (2002). Betriebssysteme - Prinzipien und Umsetzung. München, Pearson Education Deutschland GmbH.

STANAG 4187 Ed. 3 (2001). Fuzing Systems - Safety Design Requirements. Standardization Agreements (STANAG), North Atlantic Treaty Organization (NATO), NATO Standardization Agency (NSA).

Storey, N. (1996). Safety-Critical Computer Systems. Harlow, Addison Wesley.

Tsumaki, T. and Y. Morisawa (2001). A Framework for Requirements Tracing using UML. Seventh Asia-Pacific Software Engineering Conference. Singapore, Republic of Singapore, Nihon Unisys.

Zypries, B. (2003). "Bundesjustizministerin Zypries anlässlich der Eröffnung der Systems 2003: Sicherheitsanforderungen sind der Motor von Innovationen." Retrieved 2007-09-28, from http:// www.bmj.bund.de/enid/0,5cfd7d706d635f6964092d09393031093a0979656172092d093230 3033093a096d6f6e7468092d093130093a095f7472636964092d09393031/Ministerin/Reden_ 129.html.

Chapter 13
Semi-Formal Modeling
of Multi-technological Systems I: UML

13.1 Overview

The unified modeling language (UML) is a semi-formal standardized specification language which is used in the software development field. Even though it is mainly used in the software domain it can be also be used in other fields of applications. Approaches for non-software domains include, for example, railway system modeling, general requirements and system engineering, supply-chain management, and enterprise and business process modeling. Therefore, UML offers profiles and extensions that enable the customization of UML to specific domains as well as its formalization to reduce and extend it to a formal graphical language, i.e., a unique representation of a formal model.

Semi-formal modeling (UML and SysML) is typically used in companies for system specification and requirements tracing; for system development, testing, verification and validation; and for the communication between different development teams, company parts, and subcontractors. Especially SysML is used for the modeling of multi-technological (multi-disciplinary) systems. As system specification languages both can be used to specify and trace safety requirements. Chapter 1 focuses on a self-consistent introduction and use of UML diagrams for the modeling of example safety requirements.

Section 13.2 lists typical components and properties of multi-technological systems. Section 13.3 gives a short summary of the history of UML. It mentions how UML was invented but also the further development of the first version of UML to today's version UML 2.5.1 (UML 2017). Section 13.4 explains where UML can be used and where not, e.g., reminding the reader that it is a modeling and not a programming language, even if it can be used to generate program structures. Section 13.5 gives an idea of the discussion of UML in the literature, in particular, in textbooks.

Section 13.6 is the main part of this chapter and treats UML diagrams. In Sects. 13.6.1 to 13.6.9, selected UML diagrams and the SysML requirement diagram are introduced. The focus is on some of the diagram types that are used in a similar way in SysML: class diagram, composite structure diagram, state diagram, and sequence diagram.

Section 13.7 applies those diagrams in examples of modeling different types of safety requirements. The examples are used for a classification of safety requirements into four types in Sect. 13.8.

The goal of this chapter is to apply certain UML diagrams and one SysML diagram in examples of verifying safety requirements. The goal is not to give a full overview of all UML diagrams. This provides examples of how to use well-established development tools for system development for safety analyses and support of implementation of safety properties. We introduce in this chapter all diagrams that are used for modeling the safety requirements.

The chapter summarized, translates, and partially paraphrases (Hänle 2007) and (Hänle and Häring 2008). This work itself is extensively based on (Rupp, Hahn et al. 2005; Kecher 2006; Oestereich 2006).

13.2   Properties (Classification) of Multi-technological Systems

We may define *multi-technological systems* as consisting of components that use different technologies:

– (Meso, micro, nano) mechanical components;
– Electromechanical motors;
– (Micro) electronics: Micro-controllers, ASICs, FPGA, Flash Memory;
– Software;
– Optical components (e.g., laser diodes);
– Pyrotechnical, energetic, explosive components; and
– Energetic materials: batteries.

Possible properties of multi-technological systems are.

– Distributed;
– Self-sustaining, self-sufficient, autarkic;
– Closed;
– Open;
– Network(ed) system;
– Adaptable;
– Static;
– Dynamic;
– Automated;
– Human-controlled;
– Real time, soft/hard real-time system (e.g., rather long versus very short response times, or times to respond);
– Time-critical, non-time-critical;

- Adaptive, non-adaptive; and
- Intelligent.

13.3 History

The concept of object orientation is known since the 1970s (Oestereich 2006) but was not popular at first. It became widely used in the 1990s when software-producing companies started to realize the advantages of object-oriented programming (Kecher 2006).

With the beginning of the increasing use of object-oriented programming, a big variety of object-oriented analysis and design methods was developed (Rupp, Hahn et al. 2005; Kecher 2006; Oestereich 2006). Among those methods were the methods by Grady Booch, James Rumbaugh, and Ivar Jacobsen (Rupp, Hahn et al. 2005; Kecher 2006) who can be seen as the founding fathers of UML (Hänle 2007). At the beginning of the 1990s, they developed, independently of each other, modeling languages which support object-oriented software development (Hänle 2007). Of those methods, the methods by Booch and Rumbaugh were preferably used (Oestereich 2006).

In the middle of the 1990s, the software company Rational Software (IBM 2012) recruited first Booch and Rumbaugh and later also Jacobsen (Rupp, Hahn et al. 2005). While working for the same company, they had the idea to unify their modeling approaches (Oestereich 2006), and hence the name "Unified Modeling Language."

Many important companies realized that the unification introduced a quasi-standard and that a uniform modeling language simplified the communication between the developers (Rupp, Hahn et al. 2005; Kecher 2006). Some of those companies, for example, IBM, Microsoft, and Oracle, founded the consortium "UML Partners" to improve the use of UML in the industry (Kecher 2006).

In 1997, UML Version 1.1 was submitted to and accepted by the OMG (Object Management Group, "an international, open membership, not-for-profit computer industry consortium [for standardization] since 1989" (Object Management Group 2012)). Since then, the further development of UML has been done by the OMG (Rupp, Hahn et al. 2005; Kecher 2006; Oestereich 2006). The following versions included some corrections and extensions of UML.

UML 2.0 from 2004 was an in-depth revision of UML. This was necessary because the corrections and extensions did not only improve UML but also made it more complex and harder to study (Kecher 2006; Oestereich 2006). Unnecessary or unused elements, for example, parameterizable collaborations, were left out and other elements, for example, appropriate notations to specify real state systems, were added (Rupp, Hahn et al. 2005). Areas with weak semantics were specified (Rupp, Hahn et al. 2005). The changes are described in more detail in (Rupp, Hahn et al. 2005).

The most recent version is UML 2.4.2. A major difference between the newer versions of UML (UML 2.2 and later versions) compared to UML 2.0 is the so-called profile diagram which did not exist in the earlier versions of UML.

13.4   Limitations and Possibilities of UML

UML is a complex and standardized modeling language with a wide application range. It is not limited to software development or certain application areas (Rupp, Hahn et al. 2005; Kecher 2006). For example, UML 2 Profile also makes it possible to modify UML for the area where it is supposed to be used.

UML is not a programming language. However, the initiatives to "Model Driven Architecture" (MDA) are close to transforming UML into executable UML (Unhelkar 2005), that is, that an executable code in a programming language is produced from UML models.

UML is not a process model for software development as, for example, the V-Model, the Waterfall Model, etc. It rather supports existing models with a standard set of diagrams and notational elements as tools for modeling, documentation, specification, and visualization of complex systems (Rupp, Hahn et al. 2005; Unhelkar 2005; Kecher 2006; Oestereich 2006). It still is no complete substitute for a textual description (Rupp, Hahn et al. 2005; Kecher 2006).

UML can be used to model any kind of software system. Its strength lies in the modeling of object-oriented programming languages but it can also be used for procedural languages (Rupp, Hahn et al. 2005; Kecher 2006) and to model real-time constraints (He and Goddard 2000).

An overview of current application domains is given in (Lucas et al. 2009) along the main focus on UML model consistency management or on UML-based model execution in (Ciccozzi et al. 2019). Due to the fast developing computer technology and the (da Silva et al. 2018) complex demands on today's software development, UML is neither perfect nor complete (Rupp, Hahn et al. 2005).

13.5   UML in the Literature

13.5.1   Scientific Activity Around UML

The literature database WTi (WTI-Frankfurt eG 2011) contains 1379 entries for the search request "UML" since 1998. Most of the recent articles do not describe the UML itself or extensions of it. They focus on applications. The articles about UML itself often treat the verification of diagrams or the automation of generating code from diagrams. A few examples of articles are listed in this section. Correspondingly, reviews tend to be domain-specific, for instance, on executable UML models (Ciccozzi et al. 2019), on application of the diagram-type activity diagram in the safety domain (Ahmad et al. 2019), on code generation from UML models (Mukhtar and Galadanc 2018), on UML-based domain-specific languages for self-learning systems (da Silva et al. 2018), or UML-based test case generation (Pahwa and Solanki 2014).

Engels et al. (2003) present an approach to model-based analysis of a UML model. The UML model is transferred to the formal language CSP (communicating sequential processes) and tested for absence of deadlocks, that is, situations where two processes wait to be authorized to access a resource occupied by the other process and where none of the two processes releases its resource, with an existing model checker. They describe how they would develop, formally verify, and test a system in UML but only refer to a small part of UML. Zheng and Bundell (2009) introduce a new method for a model-based software component test in UML.

Elamkulam et al. (2007) use similarities between hardware circuits and embedded systems to test state machine diagrams of an embedded system using the tool IBM RuleBase which normally is a software to test hardware modules. For this, they developed a tool to transfer state machine diagrams into the formal language "Property Specification Language" which enables testing the diagrams with the tool.

Fontan et al. (2007) combine UML with expressions of a temporal logic called Allen's temporal logic. This makes it possible to verify chronological orders. Temporal logic can, in particular, be used to assess software safety of real-time embedded software (Larisch et al. 2011). Ober et al. (2006) present a method and a tool to simulate and validate UML methods which were supplemented by temporal expressions. In such approaches, UML must be restricted and extended to be able to represent a formal model.

USE is a program to validate UML diagrams which was developed in projects, theses, and dissertations at the university of Bremen (Gogolla, Büttner et al. 2006).

Wagner and Jürjens (2005) present an approach to use UML to identify classes which are prone to errors. For this, they apply a complexity analysis. The analysis uses a version of the cyclomatic complexity, a metric introduced by McCabe (1976) to measure the complexity of a program. Their version is adapted to the model level.

Ziadi et al. (2011) describe "an approach for the reverse engineering of Sequence Diagrams from the analysis of execution traces produced dynamically by an object-oriented application."

Vogel-Heuser et al. (2011) show that "object-oriented model based design can beneficially be applied in industry and that the code automatically derived from the UML model can be implemented on industrial PLCs (Programmable Logic Controllers) without additional effort."

Soeken et al. (2010) did research on the verification of UML models. They "describe how the respective components of a verification problem, namely system states of a UML model, OCL constraints, and the actual verification task, can be encoded and afterwards automatically solved using an off-the-shelf SAT solver."

13.5.2 Standard Books

We will only mention some diagrams here. For a more detailed introduction, please refer to the English book (Fowler 2003), the German books (Rupp, Hahn et al. 2005; Oestereich 2009; Kecher 2011), and the UML specification (Object Management Group 2011).

If one wants to study UML in more detail or wants to study one specific diagram more intensely, the book (Rupp, Hahn et al. 2005) can be suggested. In this book, the diagrams are described comprehensively and detailed (Hänle 2007). The book can be used by beginners as a study guide or by advanced users as reference (Hänle 2007). Since the book is from 2005, it does not cover the newest version of UML and does not include the profile diagram.

Oestereich (2009) only describes the diagrams that he considers the most important and most commonly used, so the UML specification (Object Management Group 2011) or an additional book, for example, (Rupp, Hahn et al. 2005), might be needed. The book, judged on the basis of an earlier edition, is good to get a quick introduction to UML (Hänle 2007).

Kecher (2006) gives a compact and comprehensive summary of UML. For each diagram type, he has a section about the most common mistakes made when applying the diagram (Hänle 2007). All diagrams are explained but not always in all details. For the details, one additionally needs (Rupp, Hahn et al. 2005) or the UML specification (Object Management Group 2011). The newest version was released in 2011 (Kecher 2011).

(Fowler 2003) is a well-known and comprehensive English introduction to UML. Since it is a compact book like (Oestereich 2009), one again additionally needs the UML specification (Object Management Group 2011). The book is also translated into German.

13.6 UML Diagrams

There are seven diagrams to represent the structure: the class diagram, the composite structure diagram, the component diagram, the deployment diagram, the object diagram, the package diagram, and the profile diagram.

There are also seven diagrams to represent the behavior: the activity diagram, the use case diagram, the interaction overview diagram, the communication diagram, the sequence diagram, the timing diagram, and the state machine diagram.

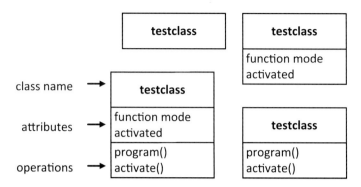

Fig. 13.1 Different forms of representing the class "testclass," following (Hänle 2007)

13.6.1 Class Diagram

A *class diagram* shows the classes used in a system and their relations where a class is an abstraction of a set of objects with the same properties (Hänle 2007).

13.6.1.1 Class

A *class* can be understood as a construction plan for objects with the same attributes and the same operations (Kecher 2006). The objects which are constructed according to the construction plan are called instances (Kecher 2006).

A class is represented by a rectangle. This can be divided into up to three segments (the class name, the attributes, and the operations), see Fig. 13.1.

The attributes describe properties of the class in form of data. For example, the attribute "activated" of Fig. 13.1 saves data in form of a truth value.

Operations describe the class's behavior. In Fig. 13.1 the objects in "testclass" can be programmed and activated.

The classes can be described in more detail by adding symbols for the *visibility* and the return type to the descriptions in the diagram, see Fig. 13.2. Visibility symbols are written in front of the attribute's/operation's name. The return type is written behind the attribute's/operation's name, separated from it by a colon. If there is no return type behind an operation, it is assumed that this operation does not return a value.

The meaning of the visibility symbols can be found in Table 13.1.

13.6.1.2 Association

The relationship between two classes is represented by a line which connects the two classes. It is called an (bidirectional) *association*. A plain line says that the classes

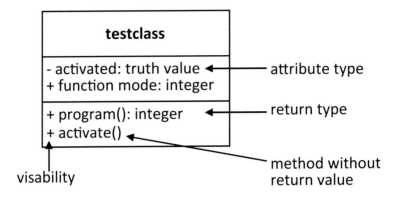

Fig. 13.2 More detailed description of the class "testclass," following (Hänle 2007)

Table 13.1 Explanation of the visibility symbols (Hänle 2007)

Symbol	Meaning	Description
+	Public	The attribute/operation is visible outside of the class. The attribute/method can be accessed from the outside and in case of an attribute its value can be changed
−	Private	The attribute/operation is not visible outside of the class. The attribute/method can only be accessed from inside the class
#	Protected	"Protected" is similar to "private" but the attribute/operation can additionally be seen and used in derived classes
~	Package	The attribute/operation is visible and accessible only for classes which are in the same package as the class the attribute/operation belongs to

know each other and can interact with each other, that is, they can access public methods and attributes of the other class (Hänle 2007).

Numbers at the ends of the line indicate the so-called multiplicity of the class. The *multiplicity of a class* is the number of instances of this class which interacts with the other class, see Fig. 13.3. (Table 13.2).

If the line between two classes is an arrow, the association is *unidirectional*, see Fig. 13.4. The class at the tip of the arrow does not know the class at the other end of the arrow and does not have access to it.

An association that links a class to itself is called a *reflexive association* (Rupp, Hahn et al. 2005; Kecher 2006), see Fig. 13.5.

Fig. 13.3 Example of an association where two instances of "testclassA" interact with one instance of "testclassB," following (Hänle 2007)

Table 13.2 Notation for the number of instances (Hänle 2007; Wikipedia 2012a)

Notation	Meaning
0.0.1	No instances or one instance
1	Exactly one instance
3	Exactly three instances
0..*	Zero or more instances
1..*	One or more instances

Fig. 13.4 Example of a unidirectional association where "testclassB" does not have access to "testclassA," following, but the reverse is true (Hänle 2007)

Fig. 13.5 Example of a reflexive association where an instance interacts with at least one (up to infinitely many) other instances of the same class, respectively (Hänle 2007)

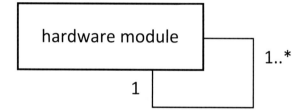

Remark (Hänle 2007): The reflexive association does not link an instance to itself. It links instances of the same class to each other.

An *aggregation* is "an association [between two classes] that represents a part-whole or part-of relationship" (Wikipedia 2012a), that is, the class on one end contains the other. The relationship is in such a way that the container and the parts can exist independently of one another. If the container is destroyed, the parts still function. It is represented by a hollow diamond shape at the end of the line where the containing class is (Wikipedia 2012a).

A *composition* describes an inseparable parts–whole relationship between the two connected classes. That is, one class, called the "part" is physically included in the other class which is called the "whole" (Rupp, Hahn et al. 2005). A part cannot exist without the whole and vice versa. The composition is described by a filled diamond shape at the end of the line where the containing class is (Hänle 2007).

13.6.1.3 Generalization

A *generalization* is a relationship that indicates that one class is a specialized form of the other. The more general class is called the superclass and the specialized class is the subclass. The relationship is represented by a line with a hallow triangle at the end of the superclass (Wikipedia 2012a).

13.6.1.4 Realization

In a *realization* relationship, the client "realizes (implements or executes) the behavior that the other model element (the supplier) specifies" (Wikipedia 2012a). The realization is represented by a dashed line with a unfilled arrow peak at the end of the supplier (Wikipedia 2012a).

13.6.1.5 Dependency

A *dependency* is weaker than an association. It only states that a class uses another class at some point in time. "One class depends on another if the latter is a parameter variable or local variable of a method of the former" (Wikipedia 2012a). It is represented by a dashed arrow from the dependent to the independent class (Wikipedia 2012a).

For an overview of all relationships between classes, see also Table 13.3.

13.6.2 Classifier

A *classifier* is a metaclass (a class where the instances are classes themselves) of a number of UML elements. It represents the common properties of the different elements (Object Management Group 2004; Hänle 2007).

Classifiers can represent different kinds of elements, for example, classes, interfaces, or associations. They are very useful for the description of the UML because they make it possible to use more abstract descriptions. They are a concept that does not visually appear in the actual UML modeling of the system (Hänle 2007).

Table 13.3 Overview of the relationships between classes

Name	Symbol	Description in words
Association	Line	Interacts with
Aggregation	Line with hollow diamond	"Has a"
Composition	Line with filled diamond	"Owns a"
Generalization	Line with hallow triangle	"Is a (type of)"
Realization	Dashed line with hallow triangle	Is a realization/implementation of
Dependency	Dashed arrow (dashed line with arrow symbol)	"Depends on"

13.6.3   Composite Structure Diagram

The *composite structure diagram* shows the interior of a classifier and its interactions with the surroundings. It presents the structure of the system and a hierarchical organization of the individual system components and their intersections (Hänle 2007).

The advantage compared to the class diagram is that in the composite structure diagram classes can be nested into each other. This describes the relations between the classes better than in the class diagram where they could only be modeled by associations (Hänle 2007).

The composite structure diagram can be seen as a concrete possible implementation of the class diagram (Hänle 2007). In a composite structure diagram, every object has a role, that is, a task in the system that it has to fulfill (Weilkiens 2006). This role level closes the gap between the class level and the object level (Weilkiens 2006).

Each object of a class can only have one role but a class can be used for several roles by assigning them to different objects of the class (Hänle 2007).

In the composite structure diagram, the following notational elements are used (see also Fig. 13.6):

- Structured classes,
- Parts,
- Connectors,
- Ports, and
- Collaborations.

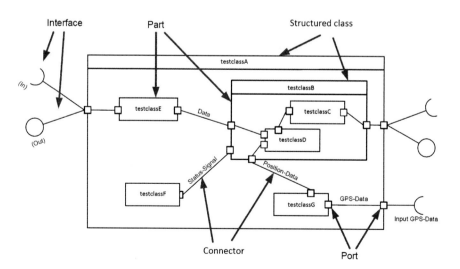

Fig. 13.6 Example of a composite structure diagram, following (Hänle 2007)

13.6.3.1 Structured Classes and Parts

A *structured class* is an alternative way to represent a class. Like in the representation of a class in the class diagram, the class is written in a horizontally divided rectangle with the name in the top segment of the rectangle. Instead of listing the attributes and operations below this as we have done it in the class diagram, the lower part is now used to display the parts of the class. This emphasizes the parts–whole relations (Hänle 2007).

A *part* of a structured class again represents a class and can again have parts. A part without its own parts is represented by a rectangle with its name. A part with own parts is represented like a structured class, only within the boundaries of another structured class (Hänle 2007).

Example In Fig. 13.6, "testclassB" is a part of "testclassA" and again contains two parts ("testclassC" and "testclassD"). "TestclassC" and "TestclassD" are only represented by rectangles because they do not contain parts while "testclassA" and "testclassB" are represented by the more complex notation for structured classes because they contain parts.

13.6.3.2 Connectors

Similarly, to associations, *connectors* represent the relationship between structured classes and/or parts. They differ from associations in representing relationships of roles rather than of general classes (Hänle 2007).

Like the association, the connector is represented by a line (see Fig. 13.6).

13.6.3.3 Ports

A *port* represents an interaction point of a classifier (Object Management Group 2004; Hänle 2007), that is, of a structured class or a part.

An interaction point is very abstract. It can specify the available interfaces of the classifiers and the services that the classifier expects (for example, from other classifiers). A port can also be used to show communication with other ports and classifiers (Hänle 2007).

A port is represented by a square which is placed on the boundary of the classifier it belongs to (Hänle 2007).

An offered or needed interface is represented by a circle or half-circle, respectively. It can be linked to a port by a line, see Fig. 13.6 (Hänle 2007).

Connectors are used to represent communication with another port or a classifier (Hänle 2007).

13.6.3.4 Collaborations

A *collaboration* describes a structure of roles which fulfill a certain function (Object Management Group 2004). These roles should be performed by instances that fulfill the function. An instance can play a role in more than one collaboration at a time (Hänle 2007).

A collaboration is represented by a dashed ellipsis. The ellipsis is divided into two parts by a straight horizontal dashed line. The upper part contains the name of the collaboration. The lower part describes the collaboration. The roles of the collaboration are defined in terms of parts and should have meaningful names. The relevant relations between roles are modeled through connectors (Hänle 2007).

Figure 13.7 shows an example of how a collaboration can be represented. The collaboration consists of the three roles RoleA, RoleB, and RoleC. RoleB interacts with both other roles. RoleA and RoleC do not interact directly with each other.

It is not declared in the representation of the collaboration in Fig. 13.7 which instances fulfill the roles of this collaboration. This can be seen in the so-called collaboration application, see Fig. 13.8.

The collaboration application is represented by a dashed ellipsis from which dashed straight lines lead to the instances that fulfill the roles of the collaboration. The name of an instance is written in a rectangle and the name of the matching role is written on the line connecting the rectangle with the ellipsis. The name of

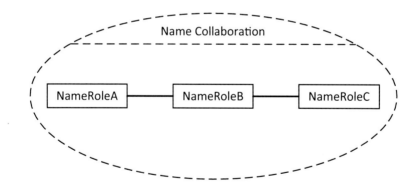

Fig. 13.7 An example of the representation of a collaboration, based on (Hänle 2007)

Fig. 13.8 Example of a collaboration application to the collaboration of Fig. 13.7, based on (Hänle 2007)

the collaboration application is written into the ellipsis, followed by a colon and the name of the collaboration.

13.6.4 State Diagram/State Machine

A *state machine* describes the behavior of a part of the system (Object Management Group 2004). This part of the system is normally a class (Weilkiens 2006) but it can also be any other classifier, for example, UseCases (Rupp, Hahn et al. 2005).

A *state diagram* describes a state machine. The elements represented in a state diagram usually are.

– states,
– transitions,
– pseudostates, and
– signals and events,

See, for example, Fig. 13.9.

State machines are based on the work of David Harel who combined the theories of general Mealy and Moore models to create a model that can be used to describe complex system behavior (Object Management Group 2004; Weilkiens 2006).

13.6.4.1 Transitions

A *transition* is a change of one state to another. It is represented by a solid arrow. The arrow can be labeled by the three elements such as trigger, guard, and behavior in the form.

trigger [guard]/behavior.

A *trigger* names an event that has to occur in order for the state machine to change to the state at the arrow peak. If there is more than one trigger, they are separated by a comma (Hänle 2007).

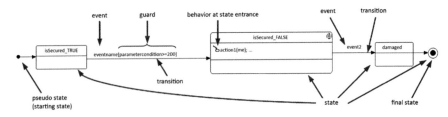

Fig. 13.9 Example of a state diagram, following (Hänle 2007)

The *guard* is a condition that has to be fulfilled for the state transition. If it is not fulfilled, the system stays in the old state. The guard is written in square brackets and can be evaluated by True or False at any point in time (Rupp, Hahn et al. 2005).

The *behavior* describes which function is executed during the state transition (Hänle 2007).

13.6.4.2   Signals and Events

Events can occur in several places in the diagram. An event can be a trigger for a state transition. It is also possible to define an event that leads to the execution of a behavior within a state. In this case, the event does not lead to a state transition but to the execution of a function (Hänle 2007).

An event is called a *signal* if it is created at and sent from a signal station (i.e., the arrow starts at a signal station) (Hänle 2007).

13.6.4.3   State

"A state models a situation during which some (usually implicit) invariant condition holds" (Object Management Group 2004).

A *state* is represented by a rounded rectangle. The rectangle is divided by a straight horizontal line. The upper part contains the name. The lower part can be used to describe the state's internal behavior. The internal behavior also has a trigger but does not lead to a state transition.

13.6.4.4   Initial State

The *initial state* is a pseudostate. It is represented by a small filled circle (Object Management Group 2004).

The initial state is defined by only one outgoing transition to the state of the machine where the process really starts. It does not have any ingoing transitions. The outgoing transition is not allowed to have a guard. This prevents situations where the state machine cannot be started. It is allowed to have a trigger and behavior on the outgoing transition but the trigger has to be an event which leads to the start of the state machine (Object Management Group 2004; Hänle 2007).

Remark Rupp, Hahn et al. (2005) suggests not to use triggers on the outgoing transition because it is normally not necessary.

13.6.4.5   Final State

A *final state* is represented by a small filled circle surrounded by another circle (Object Management Group 2004).

There can be more than one final state in a state machine. A final state is a state the machine adopts when it has completed its task. A final state is not allowed to have outgoing transitions. Ingoing transitions can have triggers, a guard, and a behavior as any normal transition.

13.6.5   Sequence Diagram

Sequence diagrams describe the temporal exchange of messages between objects (Hänle 2007). The diagram is, however, exemplary and concrete and with this generally not always complete (Oestereich 2006).

The most important notational elements in a sequence diagram are lifelines and messages.

A *lifeline* represents a participant in the interaction. The lifeline consists of a rectangle containing name and/or type of the participant and of a dashed vertical line on which the chronological sequence of messages is displayed. For an exemplary lifeline, see Fig. 13.10. In the diagram, time progresses from top to bottom. Lifelines are arranged next to each other (Hänle 2007).

A *message* represents communication between participants of the interaction. A message can be an operation call or a sent signal. A message is represented by an arrow from one lifeline to another. An answer is represented by a dashed arrow (Hänle 2007).

One can distinguish between synchronous and asynchronous calls by using full heads (for synchronous calls) and stick heads (for asynchronous calls) on the arrows (Wikipedia 2012b).

Fig. 13.10 An exemplary display of a lifeline

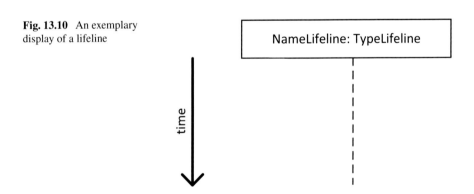

"A message sent from outside the diagram can be represented by [an arrow] originating from a filled-in circle ('found message' in UML) or from a border of the sequence diagram ('gate' in UML)" (Wikipedia 2012b).

Sequence diagrams can become complex when there are many interactions. Chapter 12 of (Rupp, Hahn et al. 2005) explains methods to modularize the diagrams to have a clearer structure.

13.6.6 Timing Diagram

Like the sequence diagram, the *timing diagram* also displays interactions between objects. It describes the temporal condition of state transitions of one or more involved objects (Oestereich 2006; Hänle 2007).

A timing diagram has a horizontal time axis. The participants of the interaction and their states are listed above each other in the diagram. An example of a timing diagram is shown in Fig. 13.11 (Hänle 2007).

The state of a participant can be seen on his lifeline which is read as follows. As long as the lifeline is a horizontal line, the participant is in the same state. A vertical or diagonal line represents a state transition (Hänle 2007).

Remark The UML specification does not specifically assign a term to the lines. Hence, different names are used in the literature for what is called lifeline here.

Messages that lead to a state transition are represented by arrows like in the sequence diagram.

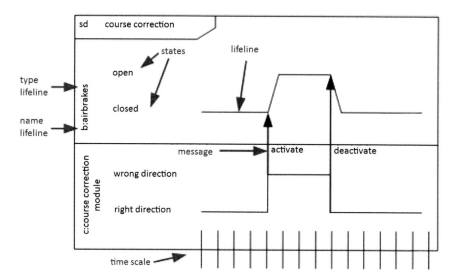

Fig. 13.11 Example of a timing diagram, following (Hänle 2007)

The time axis at the bottom of the diagram can be labeled with actual time units but it can also be kept abstract. If it is abstract, the time unit can be written in a note, see, for example, Fig. 13.26. Time points can be defined by assigning a time to a parameter, for example, $t = 10 : 56$. Time intervals can be described by two time points in curly braces, for example, $\{10 : 56..11 : 30\}$. Relations in time can be described through the parameter, for example, $t + 3$ which would be $10 : 59$ for the previous example (Hänle 2007).

In order to have the option of variable time points, the keyword "now" is available in UML (Object Management Group 2004). This keyword always refers to the current time. From this, time dependencies can be named, see, for example, Fig. 13.26 (Hänle 2007).

Another keyword of UML is "duration" to determine time spans that can be used for further tasks (Hänle 2007).

13.6.7 Further UML Diagrams

This subsection briefly introduces further UML diagrams. For a deeper introduction, please refer to the literature, for example, the books presented in 1.5.

13.6.7.1 Component Diagram

A *component diagram* in the unified modeling language depicts how components are wired together to form larger components or software systems.

13.6.7.2 Deployment Diagram

A *deployment diagram* in the unified modeling language serves to model the physical deployment of artifacts on deployment targets.

13.6.7.3 Object Diagram

An *object diagram* in the unified modeling language is a diagram which shows a complete or partial view of the structure of a modeled system at a specific time.

13.6.7.4 Package Diagram

A *package diagram* in the unified modeling language depicts the dependencies between the packages that make up a model.

13.6.7.5 Activity Diagram

Activity diagrams are, loosely defined, diagrams for showing workflows of stepwise activities and actions, where one can, for example, display choice, iteration, and concurrency. In the unified modeling language, activity diagrams can be used to describe the business and operational step-by-step workflows of components in a system. An activity diagram shows the overall flow of control.

13.6.7.6 Use Case Diagram

In software engineering, a *use case diagram* in the unified modeling language is a type of behavioral diagram defined by and created from a use case analysis. Its purpose is to present a graphical overview of the functionality provided by a system in terms of actors, their goals (represented as use cases), and any dependencies between those use cases.

13.6.7.7 Interaction Overview Diagram

Interaction overview diagrams in the unified modeling language are a type of activity diagram. They are a high-level structuring mechanism for sequence diagrams.

13.6.7.8 Communication Diagram

A *communication diagram* models the interactions between objects or parts in terms of sequenced messages. Communication diagrams represent a combination of information taken from class, sequence, and use case diagrams describing both the static structure and dynamic behavior of a system.

13.6.8 Profiles

"A key strength of UML is its ability to be extended with domain-specific customizations—the so-called profiles" (Lavagno and Mueller 2006).

Profiles make it possible to adapt UML to specialized fields of application. Within a profile, so-called "stereotypes" and "constraints" can be defined. They lead to a more detailed description of certain elements in the model. This improves the comprehensibility of the modeling by explaining aim and role of the model elements in more detail (Kecher 2006).

Profiles can only be used to extend UML descriptions. They cannot be used to change or delete parts of the modeling language (Hänle 2007).

13.6.8.1 Constraints

A *constraint* defines restrictions for the whole model. They are written in curly braces and have to be evaluable in terms of a truth value. A constraint can be written in programming code or in natural language. An example for the first is $\{size > 5\}$ while an example for the second is {A class can have at most one upper class} (Hänle 2007).

A tool-based verification of constraints is only possible if they were defined in a formal language (Object Management Group 2004) and if the UML tool in use supports this function (Hänle 2007).

13.6.8.2 Stereotypes

A *stereotype* extends the description of an element in the model. In this way, the elements can be defined and labeled more precisely (Hänle 2007).

Example (Hänle 2007): A class that describes hardware and software components could use the stereotypes «hardware» and «software» to distinguish the two types of component.

A stereotype is defined by a rectangle which contains the keyword "stereotype" and the name of the stereotype. The association of a stereotype to a metaclass whose properties the stereotype should have is indicated by an arrow. The tip points to the metaclass. As described for classes in Sect. 13.6.1.1, a stereotype can have attributes, in this context called "tagged values" (Hänle 2007).

In a UML model, a stereotype is represented by its name which is written in guillemet pointing to the outside (e.g., «hardware»). The stereotype is written in front of or above the name of the element to which this stereotype is assigned. If a stereotype has tagged values, these are written on a schematic note which is connected to the element of the model by a dashed line. The element of the model is extended by these attributes (Hänle 2007).

13.6.9 SysML Requirement Diagram

UML does not have a diagram to describe requirements. This can be found in the system modeling language (SysML), the so-called *requirement diagram*. It shows all requirements and their relations with each other and with other elements of the model (Hänle 2007). SysML will described in detail in Chap. 14.

The requirements are described with the same notation as classes, extended by the stereotype «requirement». They are only allowed to have two attributes, "text" and "id" (Weilkiens 2006).

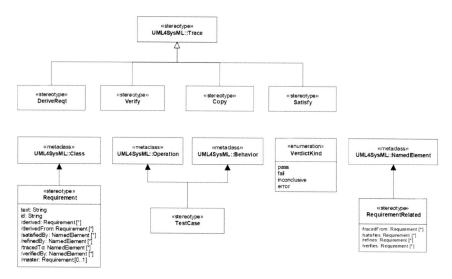

Fig. 13.12 Abstract syntax for requirements stereotypes (OMG 2010)

The attributes "text" and "id" are simple strings that the modeling person can define. The attribute "id" should be a unique ID for the requirement; the attribute "text" should describe the requirement. For an example, see Fig. 13.24 (Hänle 2007).

Figure 13.12 shows the abstract syntax for requirements stereotypes as presented in the SysML specification (OMG 2010).

13.6.9.1   Specialization of the «requirement» Stereotype

Weilkiens (2006) suggests that one uses specialized «requirement» stereotypes for specific project because SysML does not categorize the requirements. Specialized «requirement» stereotypes can distinguish between different categories of requirements. For example, safety requirements can be labeled by the stereotype «safetyRequirement» (Hänle 2007).

13.6.9.2   Relationships

SysML includes several *relationships* between requirements and other elements of the diagram. For a complete list, please refer to the SysML specification (OMG 2010). We explain the four relationships satisfy, namespace containment, refine, and trace here.

The *satisfy relationship* says that a design element fulfills a requirement (Weilkiens 2006). This is modeled by a dashed arrow with the stereotype «satisfy» written on it, see Fig. 13.13 (Hänle 2007).

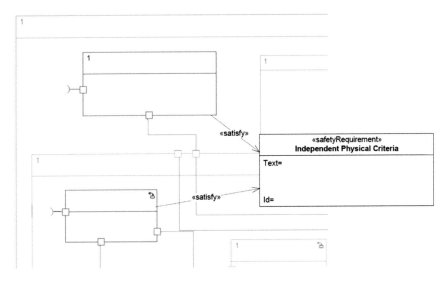

Fig. 13.13 Example of a satisfy relationship, following (Hänle 2007)

It is also possible to denote the satisfy relationship on a note. Here, the note contains the word "Satisfies" in bold letters in the headline and below the stereotype «requirement» followed by the requirement. The note is attached to the design element by a dashed line, see Fig. 13.14 (Hänle 2007).

The *namespace containment relationship* describes that one requirement is contained in another requirement (Weilkiens 2006). This relationship is represented by a line from the contained requirement to the containing requirement with a circle with a cross at the end of the containing requirement, see Fig. 13.15 (Hänle 2007).

The *refine relationship* describes that an element of the model describes the properties of a requirement in more detail (Weilkiens 2006). It is modeled in the same way

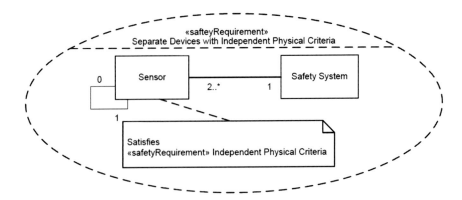

Fig. 13.14 Example of a safety relationship displayed in a note, extract from Fig. 13.23

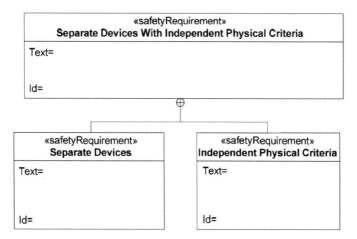

Fig. 13.15 Example of a namespace containment relationship, extract from Fig. 13.23

as the satisfy relationship, only with the stereotype «refine» on the arrow instead of «satisfy». In the alternative notation, "Satisfies" is replaced by "RefinedBy" (Hänle 2007).

The *trace relationship* is a general relationship that only states that two elements are related without further specification. It is also modeled in the same way as the satisfy relationship, only with the stereotype «trace» on the arrow instead of «satisfy». In the alternative notation, "Satisfies" is replaced by "TraceFrom" or "TraceTo" (Hänle 2007).

13.7 Example Diagrams Used for Safety Requirements

13.7.1 Example Diagrams for Single Device

In Chap. 12 several properties of safety requirements were introduced. An example of a non-functional, active, concrete, technical, and static safety requirement that can be defined for a safety–critical system is that all elements of a safety system should be located in a single safety device (Hänle 2007).

We want to use different levels of structuring to describe this safety requirement. A similar method is described in (Rupp, Hahn et al. 2005) but they use a use case diagram in the first step. Use case diagrams show the most important functions of the system and the interactions with actors outside of the system (for more details on use case diagrams, please refer to the literature or the UML specification). This is not appropriate for our example because use case diagrams should be used for functional safety requirements and we deal with a non-functional requirement (Hänle 2007).

Fig. 13.16 Safety
requirement "single device"
in a SysML requirement
diagram, following (Hänle
2007)

«safetyRequirement» **Single Device**
Text= "The components of a safety system should be located in a single device." Id=   "1"

Instead of a use case diagram, we use a SysML requirement diagram to display the safety requirement, see Fig. 13.16. We define the stereotype "safetyRequirement" as a specialization of the usual "requirement" stereotype. For simplicity, we use the ID "1" (Hänle 2007).

The next structural level is visualization through a UML collaboration. Regarding the requirement one notices that there are two types of elements which play an important role: the whole safety system and the separate elements of the safety system.

The distribution of roles is an important aspect of safety requirements. In UML, roles can be represented by collaborations in the composite structure diagram, see Sect. 13.6.3.4.

Figure 13.17 shows the collaboration for our safety requirement. Apart from the newly defined stereotype "safetyRequirement" only standard notation is used. The collaboration states exactly what the requirement asks. It says that it is a parts–whole relation and that all elements with the role "Element of Safety System" have to be located within the safety system.

We applied the safety requirement to a system from a project of the Fraunhofer Ernst-Mach-Institut. The system under regard with erased component names is displayed in Fig. 13.18. In the figure, new names are assigned to some of the components to be able to refer to them. For example, the safety system is called "deviceA" (see Fig. 13.18).

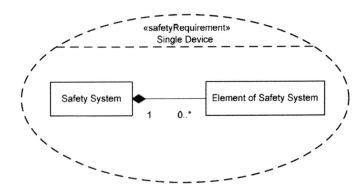

Fig. 13.17 Representation of the safety requirement "single device" in a collaboration (Hänle 2007)

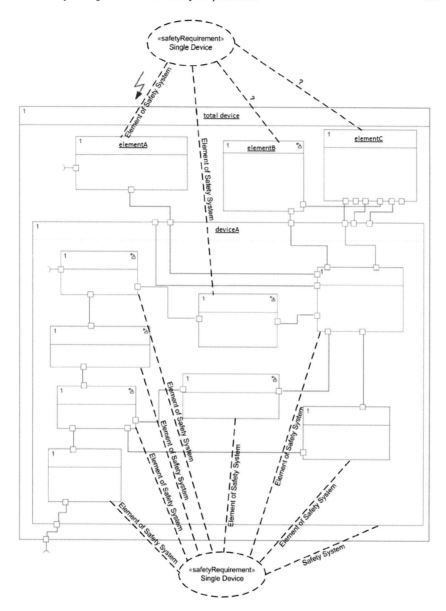

Fig. 13.18 The safety requirement collaboration applied to an example system, following (Hänle 2007)

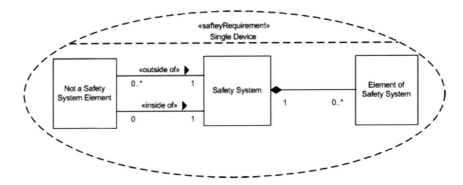

Fig. 13.19 Extension of the collaboration of Fig. 13.17 (Hänle 2007)

For every module, it had to be decided which role in the collaboration it has. For this, it was important that a person with substantial knowledge of the system was involved (Hänle 2007).

Remark A UML simulation can also help to identify the roles of the components (Storey 1996). This option was also used for the system of the project.

It turned out that one element ("elementA" in Fig. 13.18) was an element of the safety system "deviceA" but was not located within "deviceA." Hence, the safety requirement was not fulfilled.

During the analysis of the system of Fig. 13.18 it turned out that there was no role that could have been assigned to "elementB" or "elementC." Hence, the collaboration was extended to the one shown in Fig. 13.19. In the extension, the new role "Not a Safety System Element" is defined. It is connected to the safety system by two associations with the new stereotypes "outside of" and "inside of" which indicate whether the element that is not part of the safety system is located within or outside of the safety system. With this extension, one can assign the role "Not a Safety System Element" to "elementB" and "elementC" which are outside of the safety system.

The whole safety requirement is summarized in Fig. 13.20.

The idea of such semi-formalized safety requirements is that they have to be fulfilled in UML models of safety–critical systems, see (Hänle and Häring 2008). We can address them as safety patterns.

13.7.2 Example Diagrams for Separate Devices

A similar analysis can be done with the safety requirement of Fig. 13.21.

The safety requirement is represented in a collaboration in Fig. 13.22. The association between the roles "Safety System" and "Sensor" says that a safety system must be connected to at least two sensors. This is not explicitly mentioned verbose

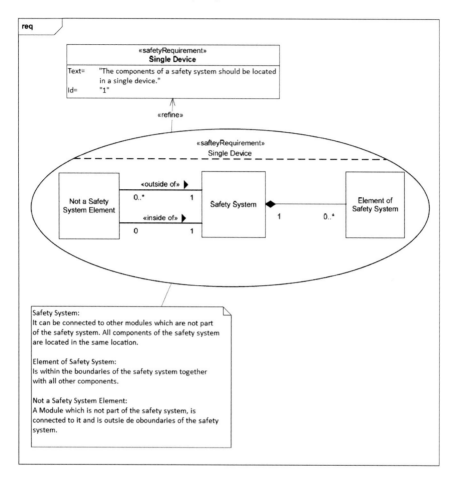

Fig. 13.20 Summary of the safety requirement "single device" (Hänle 2007)

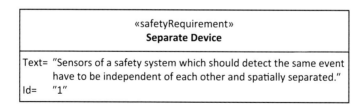

Fig. 13.21 SysML requirement diagram for the safety requirement "separate devices," following (Hänle 2007)

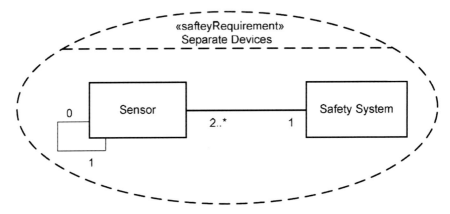

Fig. 13.22 Collaboration representing the safety requirement "separate devices" (Hänle 2007)

in the safety requirement but follows from it, see the symbol "2..*". The association does not say whether the sensors are within or outside of the safety system. The association of the role "Sensor" to the role "Sensor" is reflexive. However, it says that no sensor ("0") is allowed to be in contact with another sensor.

The safety requirement can be applied to projects in a similar way as the safety requirement in Fig. 13.18.

13.7.3 Example Diagrams for Separate Devices with Independent Physical Criteria

The safety requirement of separate devices can be extended by requiring not using redundant identical sensors. The new additional safety requirement is a non-functional, static, non-technical safety requirement saying that the sensors have to detect in different ways, for example, by analyzing different physical criteria. A description of the safety requirement is shown in Fig. 13.23 (Hänle 2007).

Remark (Hänle 2007): It is difficult to describe the non-technical safety requirement "independent physical criteria" in a UML diagram. It is, however, possible to include it in the collaboration of a technical safety requirement. Hence, the safety requirements "Separate Devices With Independent Physical Criteria" (id = "1") and "Separate Devices" (id = "1.1") are represented in a UML collaboration while "Independent Physical Criteria" (id = "1.2") is part of the complete description in Fig. 13.23 but not represented in a UML collaboration.

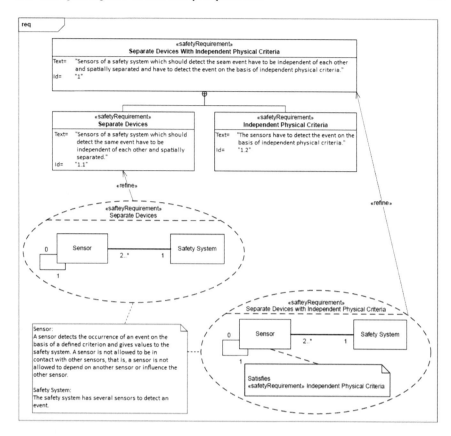

Fig. 13.23 Complete description of the safety requirement "separate devices with independent physical criteria," following (Hänle 2007)

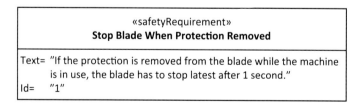

Fig. 13.24 Safety requirement "Stop Blade When Protection Removed," following (Hänle 2007)

13.7.4 Example Diagrams for a Bread Cutter

Next, we give an example of a functional, passive, and dynamic safety requirement. For this, we regard a bread cutter. The time-critical safety requirement is introduced in Fig. 13.24 (Hänle 2007).

The roles that are described in this safety requirement are the blade and the protection of the blade. "Removing the protection" and the time period of 1 s (which is a sample time only) are also important but cannot be represented in a collaboration. The safety requirement does not say how the two roles "blade" and "protection" are related to each other and to the bread cutter. In the collaboration, they are hence presented as parts of the bread cutter (Hänle 2007). (Fig. 13.25).

The requirements that could not be represented in the collaboration (the time period of 1 s and the removal of the protection) can be shown in a timing diagram. This is done in Fig. 13.26 (Hänle 2007).

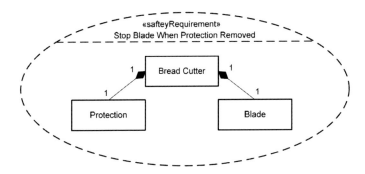

Fig. 13.25 Collaboration "stop blade when protection removed" (Hänle 2007)

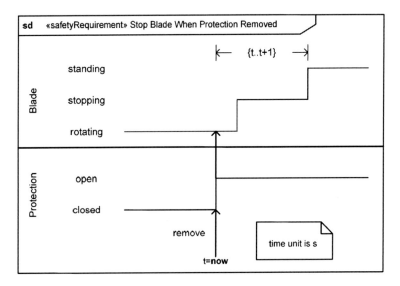

Fig. 13.26 Timing diagram for the safety requirement "stop blade when protection removed" (Hänle 2007)

The event of removing the protection and the resulting state transition of the role "protection" are indicated by the word "remove" and the vertical line from "protection closed" to "protection open." The arrow from "protection open" to "blade rotating" indicates the function that stops the blade. The parameter t is set to the current time ($t = $ now) and can be used to express the time period of "at most 1 s" (displayed as $\{t..t + 1\}$) (Hänle 2007).

In order to verify the requirement, the UML model can be simulated. The sequence diagram from the simulation can be used to test whether the removal of the protection leads to a stop of the rotating blade. If the model behaves in the way it should, the time between the occurrence of the event and the stop of the blade has to be measured in a second step, for example, by adding constructs to measure the time to the model. Both steps together would show the fulfillment of the requirements (Hänle 2007).

The complete description of the requirement can be found in Fig. 13.27.

This type of requirement is dynamic. That is, this technique only has a testing character because only concrete situations can be regarded. Hence, this method can only show the presence of failures but not their absence (Liggesmeyer 2002).

13.8   Types of Safety Requirements

Regarding the examples of Sect. 13.7, one can assign four types of safety requirements to those examples (Hänle 2007; Hänle and Häring 2008):

(1) Construction requirement,
(2) Time requirement,
(3) Requirement of a system's property, and
(4) General safety requirement.

Of course, this list is not complete. Further examples expected to yield more types of safety requirements that can be modeled with UML/SysML. For details on the four types, see Tables 13.4 and 13.5.

13.9   Questions

(1) Which types of diagrams are used in UML for basic system modeling?
(2) Name sample diagrams that can be used to graphically represent safety requirements.
(3) Who can see which part of the following class? (Fig. 13.28)

Connect the attribute/function with the matching visibility.
Activated—online inside + derived classes,
Active instances—online inside,
Activate—classes in the same package, and.

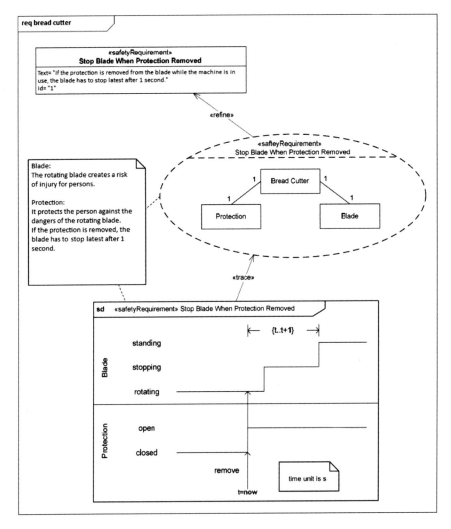

Fig. 13.27 Complete description of the safety requirement "stop blade when protection removed," following (Hänle 2007)

Nothing—outside of class.

(4) Symbolize the following relations between classes A and B:

    (a) B does not have access to A,

    (b) B is a part of A,

    (c) As composition: B is a part of A,

    (d) As realization: A is the client of B,

    (e) B is the superclass of A, and

    (f) A depends on B.

Table 13.4 Four types of safety requirements with examples. The descriptions in the left column are quoted from (Hänle and Häring 2008), the examples are assigned as in (Hänle 2007) and (Hänle and Häring 2008)

Type of safety requirement	Example
Type 1: "Construction requirement"	Sections 13.7.1 and 13.7.2
Type 2: "Time Requirement or chronological order of functions, states or events"	Section 13.7.3
Type 3: "Property of the system that has to be fulfilled and must be verifiable at all time"	Section 13.7.2, compliance with a norm
Type 4: "Requirement that describes the safety of the system, e.g., prevention of a hazard or risk"	"The system should be save" (Hänle and Häring 2008), "The system should prevent the occurrence of the hazardous event X" (Hänle and Häring 2008)

(5) What is the difference between a connector and an association?

(6) Which terms belong to which diagram?

> Terms: class, reflexive association, realization, connector, port, structured class, collaboration, classifier, profile, constraint, stereotype, event, pseudo state, trigger, message 2x, lifeline 2x, state transition, satisfy relationship. Diagrams: Class diagram, composite structure diagram, All/None, state diagram, sequence diagram, timing diagram, SysML requirement diagram.

(7) What would be a correct notation for a transition to describe the situation that an airbag is released after a shock but only if the car slows down by more than p_0 percent of its speed within time t_0?

13.10 Answers

(1) See Sects. 13.6.1, 13.6.3, 13.6.4, 13.6.7.5, 13.6.5, and 13.6.7.6.

(2) See, for instance, the diagrams used in Sect. 13.7.

(3) Activated—online inside + derived classes,

> Active instances—online inside,
> Activate—classes in the same package, and.
> Nothing—outside of class.

(4) Use Table 13.3.

(5) See Sect. 13.6.3.2.

(6) Class diagram: class, reflexive association, realization;

> Composite structure diagram: connector, port, structured class, collaboration;
> All/none: classifier, profile, constraint, stereotype;
> State diagram: event, pseudo state, trigger;
> Sequence diagram: message, lifeline;

Table 13.5 Four types of safety requirements and their representation and verification with UML/SysML (Hänle and Häring 2008)

Type of safety requirement	Required UML/SysML elements	Approach in short form
Type 1 Properties: non-functional, static, concrete, effect-oriented, technical, time-critical, (standardized) Description: Construction requirement	UML collaboration UML composite structure diagram UML/SysML sequence diagram SysML requirement diagram	1. Transform safety requirement into a requirement diagram 2. Identify and define existing roles 3. Describe the roles in a collaboration 4. Verification with»CollaborationUse« in model of the system under control 5. By simulating the system, it is possible to get a better understanding about modules which could not be assigned to a role
Type 2 Properties: functional, dynamical, effect-oriented, non-technical, (time-critical) Description: Time Requirement or chronological order of functions, states or events	UML collaboration UML timing diagram UML/SysML sequence diagram	1. Transform safety requirement into a requirement diagram 2. Identify and define existing roles 3. Describe the roles in a collaboration 4. Show the time behavior in a timing diagram 5. Verification or at least an educated guess can be achieved by simulating the system's UML model
Type 3 Properties: static, effect-oriented, non-technical, time-uncritical, (standardized) Description: Property of the system that has to be fulfilled and must be verifiable at all times	UML/SysML satisfy relationship	1. Transform safety requirement into a requirement diagram 2. Verification by analyzing verbal safety requirement with already existing methods. 3. Verification should be possible all the time (static/dynamic) 4. Only the conformance with the satisfy relation can be noted 5. Alternatively, the safety requirement can be assigned to another safety requirement that is describable in UML and in the same context

(continued)

Table 13.5 (continued)

Type of safety requirement	Required UML/SysML elements	Approach in short form
Type 4 Properties: non-functional, dynamical, abstract, cause-oriented, time-uncritical, non-technical, qualitative Description: Requirement that describes the safety of the system, e.g., prevention of a hazard or risk	UML/SysML satisfy relationship	Method as described for Type 3. Only the verification could be more demanding because the safety requirement cannot be verified all the time, only in specific situations

Fig. 13.28 Example for the visibility settings of a class

testclass
activated: truth value
- active instances: integer
~ activate()

Timing diagram: message, lifeline, state transition; and.
SysML requirement diagram: satisfy relationship.

(7) Use a transition diagram, see Sect. 13.6.4.
Use as Event 1 "airbag not released," as Trigger "shock," as Guard "v decreases to p_0 v in t_0," as Behavior "airbag_opens" and as Event 2 "airbag is released."

References

Ahmad, Tanwir; Iqbal, Junaid; Ashraf, Adnan; Truscan, Dragos; Porres, Ivan (2019): Model-based testing using UML activity diagrams: A systematic mapping study. In *Computer Science Review* 33, pp. 98–112. https://doi.org/10.1016/j.cosrev.2019.07.001.

Ciccozzi, Federico; Malavolta, Ivano; Selic, Bran (2019): Execution of UML models: a systematic review of research and practice. In *Softw Syst Model* 18 (3), pp. 2313–2360. https://doi.org/10.1007/s10270-018-0675-4.

da Silva, João Pablo S.; Ecar, Miguel; Pimenta, Marcelo S.; Guedes, Gilleanes T. A.; Franz, Luiz Paulo; Marchezan, Luciano (2018): A systematic literature review of UML-based domain-specific modeling languages for self-adaptive systems. In: 2018 ACM/IEEE 13th International Symposium on Software Engineering for Adaptive and Self-Managing Systems. SEAMS 2018: 28-29 May 2018, Gothenburg, Sweden: proceedings. With assistance of Jesper Andersson. SEAMS; Association for Computing Machinery; Institute of Electrical and Electronics Engineers; ACM/IEEE International Symposium on Software Engineering for Adaptive and Self-Managing Systems; IEEE/ACM International Symposium on Software Engineering for Adaptive and Self-Managing Systems; International Symposium on Software Engineering for Adaptive and Self-Managing Systems; International Conference on Software Engineering (ICSE). Piscataway, NJ: IEEE, pp. 87–93.

Elamkulam, Janees; Glazberg, Ziv; Rabinovitz, Ishai; Kowlali, Gururaja; Gupta, Satish Chandra; Kohli, Sandeep et al. (2007): Detecting Design Flaws in UML State Charts for Embedded Software. In Eyal Bin, Avi Ziv, Shmuel Ur (Eds.): Hardware and software, verification and testing. Second International Haifa Verification Conference, HVC 2006, Haifa, Israel, October 23-26, 2006; revised selected papers, vol. 4383. Berlin: Springer (Lecture Notes in Computer Science, 4383), pp. 109–121.

Engels, Gregor; Küster, Jochen M.; Heckel, Reiko; Lohmann, Marc (2003): Model-Based Verification and Validation of Properties. In *Electronic Notes in Theoretical Computer Science* 82 (7), pp. 133–150. https://doi.org/10.1016/s1571-0661(04)80752-7.

Fontan, B.; Apvrille, L.; Saqui-Sannes, P. de; Courtiat, J.-P. (2007): Real-Time and Embedded System Verification Based on Formal Requirements. In: International Symposium on Industrial Embedded Systems, 2006. IES '06; Antibes Juan-Les-Pins, France, 18–20 Oct. 2006. Institute of Electrical and Electronics Engineers; International Symposium on Industrial Embedded Systems; IES 2006. Piscataway, NJ: IEEE Service Center, pp. 1–10.

Fowler, M. (2003). UML Distilled: A Brief Guide to the Standard Object Modeling Language, Addison-Wesley Professional.

Gogolla, M., F. Büttner and M. Richters (2006). USE: A UML-based Specification Environment. Bremen, Universität Bremen.

Hänle, A. (2007): Modellierung und Spezifikation von Anforderungen eines sicherheitskritischen Systems mit UML, Modeling and Specification of Requirements of a safety critical System with UML. Hochschule Konstanz für Technik, Wirtschaft und Gestaltung (HTWG), University of Applied Sciences; Fraunhofer EMI, Efringen-Kirchen. Computer Science.

Hänle, A.; Häring, I. (2008): UML safety requirement specification and verification. In Sebastian Martorell, C. Guedes Soares, Julie Barett (Eds.): Safety Reliablity and Risk Analysis: Theory, Methods and Applications, European Safety and Reliablity Conference (ESREL) 2008, vol. 2. Valencia, Spain: Taylor and Franzis Group, London, pp. 1555–1563.

He, Weiguo; Goddard, S. (2000): Capturing an application's temporal properties with UML for Real-Time. In: Proceedings, Fifth IEEE International Symposium on High Assurance Systems Engineering (HASE 2000). November 15–17, 2000, Albuquerque, New Mexico. IEEE International High-Assurance Systems Engineering Symposium; IEEE Computer Society. Los Alamitos, Calif: IEEE Computer Society, pp. 65–74.

IBM. (2012). "Rational Software." Retrieved 2012-04-24, from http://www-01.ibm.com/software/de/rational/.

Kecher, C. (2006). UML 2.0: Das umfassende Handbuch, Galileo Computing.

Kecher, C. (2011). UML 2.0: Das umfassende Handbuch, Galileo Computing.

Larisch, M; Siebold, U; Häring, I (2011): Safety aspects of generic real-time embedded software model checking in the fuzing domain. In Christophe Berenguer, Antoine Grall, Carlos Guedes Soares (Eds.): Advances in Safety, Reliability and Risk Management. Esrel 2011. 1st ed. Baton Rouge: Chapman and Hall/CRC, pp. 2678–2684.

Lavagno, L. and W. Mueller. (2006). "UML: A Next-Generation Language for SoC Design." Retrieved 2012-05-16, from http://electronicdesign.com/Articles/Index.cfm?ArticleID=12552&pg=1.

Liggesmeyer, P. (2002). Software-Qualität: Testen, Analysieren und Verifizieren von Software. Heidelberg, Berlin, Spektrum Akademischer Verlag.

Lucas, Francisco J.; Molina, Fernando; Toval, Ambrosio (2009): A systematic review of UML model consistency management. In *Information and Software technology* 51 (12), pp. 1631–1645. https://doi.org/10.1016/j.infsof.2009.04.009.

McCabe, J. (1976). "A Complexity Measure." IEE Transactions on Software Engineering **SE-2**.

Mukhtar, M I; Galadanc, B S (2018): Automatic code generation from UML diagrams: the state of the art 13 (4), pp. 47–60.

Ober, I., S. Graf and I. Ober (2006). "Validating timed UML models by simulation and verification." International Journal on Software Tools for Technology Transfer (STTT) **8**(2): 128.

Object Management Group (2004). Unified Modeling Language Superstructure.

Object Management Group (2011). Unified Modeling Language 2.4.1.

Object Management Group. (2012). "OMG." Retrieved 2012-04-24, from http://www.omg.org/.

Oestereich, B. (2006). Analyse und Design mit UML 2.1. München, Oldenbourg Wissenschaftsverlag GmbH.

Oestereich, B. (2009). Analyse und Design mit UML 2.3: Objektorientierte Softwareentwicklung, Oldenbourg Wissenschaftsverlag.

OMG (2010). OMG Systems Modeling Language (OMG SysML), V1.2 - OMG Available Specification, Object Management Group.

Pahwa, Neha; Solanki, Kamna (2014): UML based Test Case Generation Methods: A Review. In *IJCA* 95 (20), pp. 1–6. https://doi.org/10.5120/16707-6859.

Rupp, C., J. Hahn, S. Queins, M. Jeckle and B. Zengler (2005). UML 2 glasklar: Praxiswissen für die UML-Modellierung und -Zertifizierung, Carl Hanser Verlag GmbH & CO. KG.

Soeken, Mathias; Wille, Robert; Kuhlmann, Mirco; Gogolla, Martin; Drechsler, Rolf (2010): Verifying UML/OCL models using Boolean satisfiability. In: Design, Automation & Test in Europe Conference & Exhibition (DATE), 2010. 8–12 March 2010, Dresden, Germany; proceedings. European Design Automation Association; Design, Automation & Test in Europe Conference & Exhibition; DATE. Piscataway, NJ: IEEE, pp. 1341–1344.

Storey, N. (1996). Safety-Critical Computer Systems. Harlow, Addison Wesley.

UML (2017): Unified Modeling Language. Version 2.5.1. Edited by Object Management Group OMG. Available online at https://www.omg.org/spec/UML, checked on 9/21/2020.

Unhelkar, B. (2005). Verification and Validation for Quality of UML 2.0 Models, John Wiley and Sons.

Vogel-Heuser, B., S. Braun, B. Kormann and D. Friedrich (2011). Implementation and evaluation of UML as modeling notation in object oriented software engineering for machine and plant automation. 18th IFAC World Congress of the International Federation of Automatic Control, Milano, Italy.

Wagner, S. and J. Jürjens (2005). Model-Based Identification of Faul-Prone Components, Technische Universität München.

Weilkiens, T. (2006). Systems Engineering mit SysML/UML. Heidelberg, dpunkt.verlag GmbH.

Wikipedia. (2012a). "Class diagram." Retrieved 2012-05-03, from http://en.wikipedia.org/wiki/Class_diagram.

Wikipedia. (2012b). "Sequence diagram." Retrieved 2012-05-08, from http://en.wikipedia.org/wiki/Sequence_diagram.

WTI-Frankfurt eG (2011). TEMA Technik und Management. Frankfurt.

Zheng, Weiqun; Bundell, Gary (2009): Model-Based Software Component Testing: A UML-Based Approach. In: International Conference on Information Systems. ICIS 2007; Montreal, Quebec, Canada, 9–12 December 2007. Association for Information Systems; International Conference on Information Systems; ICIS. Red Hook, NY: Curran, pp. 891–899.

Ziadi, T., M. A. A. Silva, L. M. Hillah and M. Ziane (2011). A Fully Dynamic Approach to the Reverse Engineering of UML Sequence Diagrams. 16th IEEE International Conference on Engineering of Complex Computer Systems. Las Vegas, NV, US: 107–116.

Chapter 14
Semi-formal Modeling of Multi-technological Systems II: SysML Beyond the Requirements Diagram

14.1 Overview

This chapter introduces the semi-formal modeling language SysML (system modeling language) of which we have already seen the requirements diagram in Sect. 13.6.9.

The introduced SysML diagrams are suited to analyze system safety, to document system knowledge, and for requirements engineering and management. As was argued to some extent already in Chap. 13 on the use of UML for safety requirements definition and tracing, SysML is a multi-domain semi-formal language for systems and requirements specification, systems engineering, and requirements tracing. It is less abstract than UML while containing further diagrams that are well suited for efficient graphical requirements modeling, in particular, the SysML requirements diagram (req).

Section 14.3 introduces the SysML block definition diagram (bdd), the SysML internal block diagram (ibd), the SysML activity diagram (act), the SysML use case diagram (uc), the SysML state machine diagram (stm), and the SysML sequence diagram (sd). For each diagram, typical applications and examples including figures are given how to use it for modeling safety requirements. Together with the SysML requirements diagram (req) they are already sufficient for modeling most types of technical safety requirements.

Before, the history of SysML and its relation to UML is briefly summarized in Sect. 14.2. The diagram types, bdd, ibd, stm, and act, are already sufficient for modeling most types of technical safety requirements when using a state machine approach that considers the structure and behavior of the socio-technical system. It is shown how this can be contextualized with the diagram types, uc and sd, to cover system application cases. The chapter provides examples mainly in the domain of embedded systems.

I. Häring, *Technical Safety, Reliability and Resilience*,
https://doi.org/10.1007/978-981-33-4272-9_14

The chapter combines and extends (Larisch et al. 2008a, b; Siebold et al. 2010a, b; Siebold and Häring 2012), which mainly focus on embedded multi-technological systems. In a similar way, it can be applied to checkpoint systems (XP-DITE 2017; TRESSPASS 2020) and automated industrial inspection systems of chemical plants (ISA4.0 2020).

14.2   History and Relation UML and SysML

"The 'System Modeling Language' (SysML) (OMG 2007) is a new powerful semi-formal standardized specification language based on the 'Unified Modelling Language' (OMG 2005) for 'System Engineering'. The first SysML version was announced by the 'Object Management Group' in September 2007. Because UML is usually used for software modeling SysML was mainly developed to be able to model hardware and software components.

SysML consists out of some UML diagrams, modified UML diagrams and two diagram types which are not included in UML, see also Fig. 14.1 and Table 14.1. Because SysML is designed for 'System Engineering' it avoids most of the concepts of object-oriented programming used in UML" (Larisch et al. 2008a), b.

14.3   Overview of Diagrams

In SysML, there are nine different diagrams which are divided into two categories, behavior diagrams and structure diagrams, with the exceptions of the requirements diagram which belongs to none of the categories, see Fig. 14.2.

To model the static aspects of a system, e.g., the architecture of a system, the structure diagrams are used. Structure diagrams are the block definition diagram,

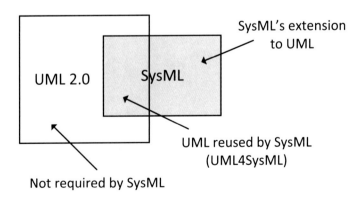

Fig. 14.1 UML-SysML comparison (Larisch et al. 2008a, b), based on (OMG 2008)

SysML diagram	Comparison to UML diagram
Activity diagram	Modified from UML activity diagram
Requirement diagram	New in SysML
Sequence diagram	Same as in UML
State machine diagram	Same as in UML
Use case diagram	Same as in UML
Block definition diagram	Modified from UML class diagram
Internal block diagram	Modified from UML composite structure diagram
Package diagram	Same as in UML
Parametric diagram	New in SysML
Not in SysML	UML deployment diagram
Not in SysML	UML communication diagram
Not in SysML	UML timing diagram
Not in SysML	UML interaction overview diagram
Not in SysML	UML profile diagram
Not in SysML	UML component diagram
Not in SysML	UML object diagram

Table 14.1 Comparison of SysML and UML diagrams, based on (Larisch et al. 2008a, b)

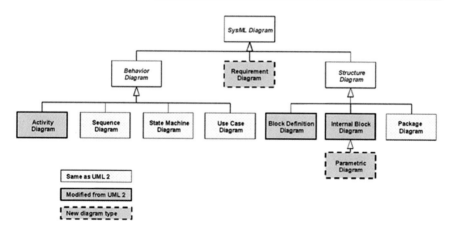

Fig. 14.2 An overview of the SysML diagrams (OMG 2014)

the internal block diagram, the parametric diagram, and the package diagram. The behavior diagrams are used to model the dynamic aspect of the system, e.g., interactions between components. Behavior diagrams are the activity diagram, the sequence diagram, the state machine diagram, and the use case diagram.

14.3.1 Block Definition Diagram

"The *Block Definition Diagram* is based on the UML Class Diagram. It describes the structure of a system […]. System decomposition can be represented by using composition relationships and associations. Therefore, so called 'blocks' are used. A block represents software, hardware, personnel facilities, or any other system element (OMG 2008). The description of a system in a Block Definition Diagram can be simple by only mentioning the names of several modules or detailed with the description of all information, features and relationships" (Larisch et al. 2008a, b).

The example in Fig. 14.3 shows all the interfaces and quantities relevant for SIL determination of safety-related systems: component type (A,B), safe failure fraction (SFF), and hardware failure tolerance (HFT), see Sect. 11.9. It has been generated by a SysML modeling environment as described in (Siebold 2013). A similar approach has also been used in (Siebold et al. 2010a, b).

The block definition diagram (bdd, BDD) shows the principal physical components of a system and its hierarchical structure in a tree-like diagram. It allows developing a number of conceptual architectures that can be used as a starting point for analysis. Block definition diagrams are based upon class diagrams. It is only meant to show composition relationships between elements and their multiplicities. Main components are called system parts or system blocks.

The block definition diagram is similar to the class diagram because it gives the allowed classes (called blocks) but does not yet show the relation between the blocks in a real system.

The bdd is well suited to provide an overview of all possible system components, their interfaces, and internal states. Often it is also used to show inclusions and part relations. For safety analysis, it is important to know all components, their internal states, and quantities that are accessible from outside.

14.3.2 Internal Block Diagram

"The internal structure of a System can be described by an *Internal Block Diagram* consisting of blocks parts, ports and connectors […]. A Part is a block inside of another block. With ports the interfaces of a block can be defined. A connector is used to connect two ports of different blocks and symbolizes a communication channel or object flow channel between these two blocks" (Larisch et al. 2008a, b).

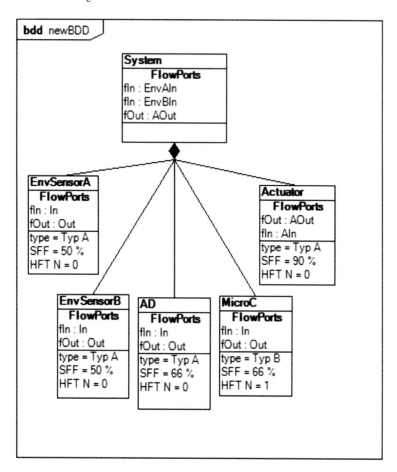

Fig. 14.3 Example for a block definition diagram (Siebold et al. 2010a, b)

An example is given in Fig. 14.4. "Two environment sensors, the instances *es:EnvSensorA* and *es:EnvSensorB*, are used to sense environment stimuli and send their measurements to the analog digital converters *ad:AD* and *ad2:AD*. The micro-controller *mc:MicroC* processes the values and can send trigger-signals to the actuator *a:Actuator*. In the sample case *Environment stimuli sensed* may mean that either of the sensors or both have sensed the stimuli. In this sense the SysML block definition and internal block diagrams explain in detail the requirement diagram" (Siebold and Häring 2012).

SysML internal block diagrams allow to identify interfaces of the systems, object flows, inclusion, and part relation of components as well as the structural (static) design. This can be used in safety analysis to determine immediate effects of events and overall system effects.

Figure 14.4 shows, for instance, the parallel structure regarding sensors and their signal evaluation as well as the linear structure for the evaluation and actuation.

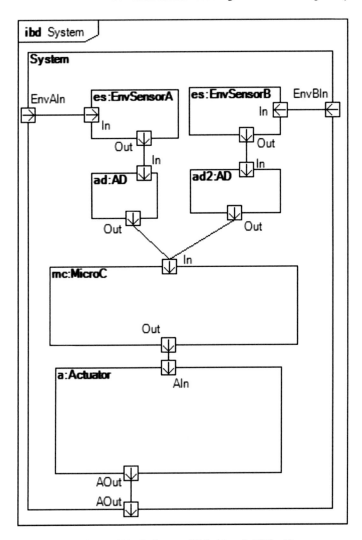

Fig. 14.4 Example for an internal block diagram (Siebold et al. 2010a, b)

14.3.3 Activity Diagram

"The *Activity Diagram* is modified from UML 2.0. With the Activity Diagram the flow of data and control between activities can be modeled. It represents the work flow of activities in a system. The sequence can be sequential or split in several concurrent sequences. Causes of conditions can be branched into different alternatives. With the use of activity diagrams one can show the behavior of a system, which is useful in the concept phase of the safety lifecycle. Duties and responsibilities of modules can be modeled […]" (Larisch et al. 2008a, b).

They are similar to flowcharts and show the flow of behavior, decision points, and operation calls throughout the model. Often used diagram elements (see Fig. 14.5) are InitialNode (or Start Node), Action, Events, DecisionNode, MergeNode, ForkNode, JoinNode, and ActivityFinal (or Final Node). The activity starts at the initial or start node and ends at the final node (OMG 2014).

Time event and accept event.

Action.

Start node.

Final node.

Decision node (if–then–else).

Merge node.

Fork node indicates parallel actions.

Join node.

SysML activity diagrams (ad) show all the allowed state transitions, the forking, and merging of the allowed state flows. This can be used to identify dynamic system behavior and to assess it regarding the implications of system and subsystem state transitions regarding safety. The activity diagram should be used after and together with the SysML state machine diagram (smd) to first identify the component, subsystem, and system states before determining all allowed transitions.

As Fig. 14.5 only shows the expected system behavior, it cannot model the effects of failures explicitly. However, it supports the inspection of system designs to determine potential failures and their effects.

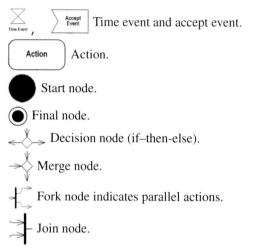

Fig. 14.5 Example of an activity diagram (Larisch et al. 2008a, b)

14.3.4 State Machine Diagram

"*State Machine Diagrams* are also adopted unmodified from UML. They describe the behavior of an object as the superposition of all possible state histories in terms of its transitions and states" (Larisch et al. 2008a, b). See Fig. 14.6 for an example.

State machine diagrams (SMD) provide means of showing state-based behavior of a use case, block, or part. Behavior is seen as a function of inputs, present states, and transitions. A transition is the path between states indicated by a trigger/action syntax.

SysML state machine diagrams determine all internal and external states of components and allowed transitions. Thus, they are well suited to assess systematically possible transitions regarding safety. Whereas activity diagrams tend to show only the intended transitions on various levels, and the state machine diagrams list all possible transitions. However, a potential drawback of the state machine diagram is that it is focusing more on single modeling elements.

14.3.5 Use Case Diagram

Use case diagrams are as in UML. Use case diagrams can be used to describe the main functions of a system, the internal processes, and the persons and systems that are connected with the system (Larisch et al. 2008a, b), see Fig. 14.7 for an example. Figure 14.8 provides a sequence diagram for the example showing a sequence of the use of the system elements in case of a collision.

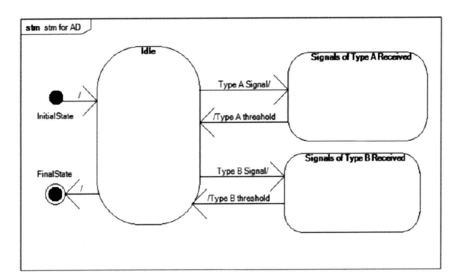

Fig. 14.6 Example of a state machine diagram (Siebold and Häring 2012)

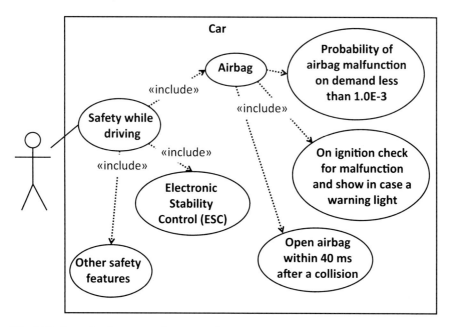

Fig. 14.7 Example of a use case diagram (Larisch et al. 2008a, b)

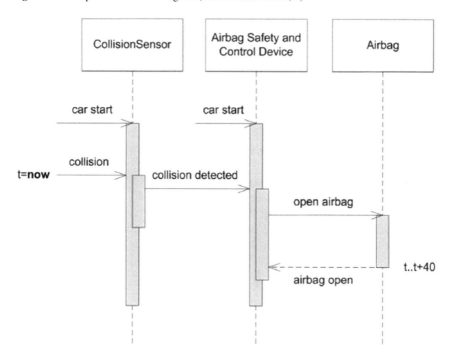

Fig. 14.8 Example of a sequence diagram (Larisch et al. 2008a, b)

Use case diagrams give the system its context. They set the system boundaries. Use case diagrams show the relationship between external actors and the expected uses of the system. They define the expected uses of the system from the point of view of the people/systems that use it. Use cases can be found from the perspective of the actors or users and from the expected system behavior, e.g., system support user 1 with three services and user 2 with four services two of which overlap.

Use case diagrams are appropriate to identify expected and potential (miss) use contexts of safety-related and safety–critical systems. They may also be used to relate the main functions of a system with its safety functions. They are versatile tools in early phases of system development to clarify the main aims and respective safety goals.

"Sequence Diagrams are adopted unmodified from UML. They describe the information flow between elements in certain scenarios […]. Sending and receiving of messages between entities are represented in sequence diagrams" (Larisch et al. 2008a, b).

Sequence diagrams (SD) are used to specify the messages that occur between model elements. They can show timing information. Interaction operators added in UML 2.0 which expand the functionality of sequence diagrams and reduce the number of scenarios needed to describe a system's behavior.

SysML sequence diagrams are of main interest for safety assessments because they show (un)intended sequencing of activities, they can be quantified using time information, and they show which components are expected to be active in which way. This information can be used to assess safety issues, for instance, all critical time requirements.

14.4 Tasks and Questions

(1) Identify the most recently released specification documents of UML and SysML and respective tutorials.
(2) Which additional semi-formal languages are currently available or under development that are conform with the UML standard?
(3) How are UML and SysML related?
(4) Name all SysML diagrams. What is, respectively, their main purpose?
(5) When in the safety life cycle would you recommend to use which types of SysML diagrams?

14.5 Answers

(1) Starting point should be the official website of the object management group (OMG) of UML and its documentation material.
(2) See answer to (1) as starting point.

(3) See Sect. 14.2.
(4) See Fig. 14.2 for overview and ordering and more detailed descriptions, in particular, in Sects. 13.6.5, 13.6.7.4, and 14.3.1 to 14.3.5.
(5) Generate a table similar to Table 8.13 by replacing the hazard analyses types with SysML diagram types and the development life cycle phases with safety life cycle phases of Fig. 11.2. For instance, SysML use case diagrams and requirement diagrams are appropriate for Phase 1.

References

ISA4.0 (2020): BMBF Verbundprojekt: Intelligentes Sensorsystem zur autonomen Überwachung von Produktionsanlagen in der Industrie 4.0 - ISA4.0, Joint research project: Intelligent sensor system for the autonomous monitoring of production systems in Industry 4.0, ISA4.0, 2020–2023.

Larisch, M., A. Hänle, I. Häring and U. Siebold (2008a). Unterstützung des Nachweises funktionaler Sicherheit nach IEC 61508 durch SysML. Dipl. Inform. (FH), HTWG-Konstanz.

Larisch, M., A. Hänle, U. Siebold and I. Häring (2008b). SysML aided functional safety assessment. Safety Reliablity and Risk Analysis: Theory, Methods and Applications, European Safety and Reliablity Conference (ESREL) 2008. S. Martorell, C. G. Soares and J. Barett. Valencia, Spanien, Taylor and Franzis Group, London. 2: 1547–1554.

OMG (2005). Unified Modeling Language, Superstructure, Object Management Group.

OMG (2007). OMG Systems Modeling Language (OMG SysML), V1.0 - OMG Available Specification, Object Management Group.

OMG. (2008). "OMG Systems Modeling Language." Retrieved 2008-02-13, from http://www.omg sysml.org/.

OMG. (2014). "OMG Systems Modeling Language." Retrieved 2014-08-07, from http://www.omg sysml.org/.

Siebold, Uli (2013): Identifikation und Analyse von sicherheitsbezogenen Komponenten in semi-formalen Modellen. Dissertation. Zugl.: Freiburg, Univ., Diss., 2013. Stuttgart: Fraunhofer Verlag (Schriftenreihe Forschungsergebnisse aus der Kurzzeitdynamik, 25).

Siebold, Uli; Häring, Ivo (2012): Semi-formal safety requirement specification using SysML state machine diagrams. In: 11th International Probabilistic Safety Assessment and Management Conference and the Annual European Safety and Reliability Conference 2012. (PSAM11 ESREL 2012); Helsinki, Finland, 25–29 June 2012. International Association for Probabilistic Safety Assessment and Management; European Safety and Reliability Association; International Probabilistic Safety Assessment and Management Conference; PSAM; Annual European Safety and Reliability Conference; ESREL. Red Hook, NY: Curran, pp. 2102–2111. Available online at http://www.proceedings.com/16286.html.

Siebold, U.; Larisch, M.; Häring, I. (2010a): Using SysML Diagrams for Safety Analysis with IEC 61508. In: Sensoren und Messsysteme. Nürnberg: VDE Verlag GmbH, pp. 737–741.

Siebold, Uli; Larisch, Matthias; Häring, Ivo (2010b): Nutzung von SysML zur Sicherheitsanalyse. In atp magazin 52 (12), pp. 54–61. Available online at http://ojs.di-verlag.de/index.php/atp_edi tion/article/view/2100.

TRESSPASS (2020): Robust risk based screening and alert system for passengers and luggage, EU Project, 2018–20121, Grant agreement ID: 787120. Available online at https://cordis.europa.eu/project/id/787120; https://www.tresspass.eu/, checked on 9/27/2020.

XP-DITE (2017): Accelerated Checkpoint Design Integration Test and Evaluation. Booklet. Available online at https://www.xp-dite.eu/dissemination/, updated on 2017-2, checked on 9/25/2020.

Chapter 15
Combination of System Analysis Methods

15.1 Overview

Chapter 2 showed for all approaches presented in the book how to potentially use them for technically driven resilience engineering of socio-technical systems. While focusing on key concepts of resilience, also neighboring concepts as classical risk analysis were used, which can be understood to be a fundamental framework for technical safety assessment. The focus was on single method assessment. Complementary to Chaps. 2 and 15 provides exemplary discussions on method combination for system reliability, safety and resilience analysis, and improvement. However, the focus of this chapter is not on details of methods and their extension options but on method combination.

The previous chapters introduced different methods of technical safety with focus on tabular and graphical system analysis methods, reliability prediction, and SysML graphical system modeling. So far, it only has been discussed in Sect. 8.2 (see, in particular, Table 8.1) and Sect. 8.11 (see Table 8.12) with focus on tabular approaches in more detail how to combine methods efficiently. Hence, this chapter provides further recommendations on method combinations.

Chapter 10 categorized and exampled development processes and Chap. 11 the sample safety development process of functional safety. Both types of processes need to be supported with appropriate techniques and measures. This is shown in more detail for methods for hardware and software in Sect. 11.8.8 by using an example of the method tables of (IEC 61508 S+ 2010) . This refers to a very elaborated approach for method selection and combination as stipulated by the generic functional safety standard and implemented in many application standards.

This chapter continues the discussion of the combination of and interaction between four different basic methods: SysML, HA, FMEA, and FTA. It uses the example of an electric vehicle and the identification of faults in after-sales scenarios. This can be understood as a use case of engineering resilience in the sense of fast stabilization, response, and recovery post potential disruptive events during operation of modern green transport systems. In particular, since electrical vehicles are strongly dependent on safe and reliable battery systems as well as operationally, economically, and technically efficient after-sales service (Automotive Word 2015; Kanat and Ebenhöch 2015; Barcin et al. 2014).

First, the chapter discusses the advantages of semi-formal modeling with SysML in combination with failure modes and effects analysis (FMEA) and fault tree analysis (FTA) in Sect. 15.2. Section 15.3 regards the connection of hazard analysis to other system analysis methods. Afterward, Sect. 15.4 discusses the combination of FMEA and FTA. Section 15.5 treats the aggregation of subsystem FTAs to a system FTA. Section 15.6 shows a way how FTA results can be used after product development to optimize error detection and repair by providing efficient failure isolation (detection) procedures also called fault isolation procedures (FIP).

In summary, the chapter shows how the classical system analysis methods work together. Overarching aim is to show that all potential hazards of systems are under control, for which a hazard analysis (HA) at system level can be used for bookkeeping.

See the following three references for more comprehensive and formal approaches. In Peeters et al. (2018), FMEA and FTA are applied recursively to an additive manufacturing system. The work in Liu et al. (2013) shows hot to combine FMEA and FTA in the case of a photovoltaic plant.

The application of a combined approach to subsea blowout events is given in Shafiee et al. (2019).

Fault tree analysis can be used to deepen the understanding of a system after an FMEA. It can find failures with more than one cause and can be used to critically reconsider the RPN in the FMEA. Both are more detailed analyses that should be supported with a graphical system model, e.g., using SysML.

A completed FTA or a part of it can be used to derive a concise fault isolation procedure for the after sales, for example, in car repair shops after the car has been sold to the customer. For this, it is useful to turn the fault tree into a template and adapt the description of the text boxes with an emphasis on the description of symptoms.

The main source of this chapter is Kanat (2014).

15.2   Conduction of SysML Modeling Before System Analysis Methods

Semi-formal modeling helps to develop, understand, document, and improve the system. SysML can, in particular, be used to model safety requirements using finite state machine modeling (Siebold and Häring 2012). Examples for modeling safety- or security-relevant systems with SysML cover a wide range, generic safety systems (Siebold et al. 2009, 2010), fuzing safety systems (Larisch et al. 2008a, b, 2009), and airport checkpoints (Jain et al. 2020; Renger et al. 2015). But it can also be applied to model such generic processes as five-step risk management (Schoppe et al. 2014), including the use of related methods (Schoppe et al. 2015).

In particular, a detailed modeled system simplifies the application of methods like FMEA and FTA. The search can be structured with the help of the SysML diagrams and can therefore lead to better and more complete results, see also (Larisch et al. 2008a; b).

There are also first attempts to automatically generate fault trees from SysML diagrams, for example, by (6-th International Disaster and Risk Conference (IDRC): Integrative Risk Management—toward resilient cities) (Li and Wang 2011) and (Mhenni et al. 2014). SysML models have also been used to automatically generate FMEA templates, e.g., (David et al. 2008). However, today, system analysis methods are typically not yet automized within practical approaches.

15.3 Combination of Hazard Analyses and Other System Analysis Methods

Chapter 8 showed the relationship between the different hazard analysis methods, see Fig. 8.1. The abbreviations used in the figure stand for hazard analysis (HA), preliminary hazard list (PHL), preliminary hazard analysis (PHA), subsystem hazard analysis (SSHA), system hazard analysis (SHA), and operating and support hazard analysis (O&SHA).

Depending on the product, the procedure of the hazard analysis can also follow different schemes, for example, the PHA and HA can be combined for very simple systems. Table 8.1 shows which form of safety analysis needs input from which other form. It reads as follows: the analysis method on the left side uses information from the methods in the columns marked with an "X".

Section 8.2 discussed how FMEA and FTA results are used to support the hazard analyses. How they are related in detail to each other is shown exemplarily in Sects. 15.4 and 15.5.

15.4 From FMEA to FTA

Fault tree analysis can be used to further analyze the results of an FMEA. In this way, the interactions of components, interfaces, and functions can be considered which are not covered in a classical FMEA, i.e., everything beyond single point of failure analysis. An FMEA also does not consider the interaction of failures and consequences.

Figure 15.1 shows that FMEA results can be used in an FTA and vice versa. The consequences from the FMEA can be used as top events for an FTA. An example of such a failure on system level is the loss of the battery system. The idea is that on system level a similar failure as caused by a single point of failure can also be

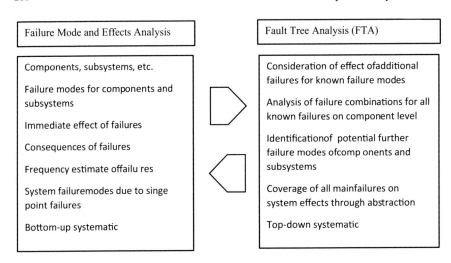

Fig. 15.1 Interaction of FMEA and FTA. Inter alia the combination of different lines of the FMEA, i.e., failure modes are considered by the FTA. The FMEA takes up all newly identified failure modes of the FTA. Based on Kanat (2014)

generated by combinations of failures. Of course, it is understood that it cannot be guaranteed that in this way all possible failures on system level can be identified.

The FTA considers interactions between components, interaction of failures, and sequences of failures. Common cause failures can be considered by having the same event in several places in the fault tree.

In the FMEA, the different failures are considered independently of each other. The FTA analyzes what happens if combinations of the failures from the FMEA occur. For example, according to the FMEA of an electrical vehicle, a shortcut leads to a destroyed main fuse. But what are the consequences of that? Then one takes advantage that there is a main contactor which disconnects the battery from the intermediate circuit in case of such a shortcut. If, however, the main contactor also does not work, the cell is damaged, especially if the battery has a low SOC. Such combinations of failures are identified by the FTA analysis.

This is an example how system failures of interest are identified by an FMEA. Then FTA is used to identify all possible failure combinations leading to the same system failures, see Fig. 15.1. At least single point of failures and faults that are found during the FTA should be included in the FMEA. Thus, both analyses are living documents and should be mutually updated if new results are found.

15.5 Combination of Component FTAs to a System FTA

Complex systems are often not analyzed in one fault tree. Again, we consider the example of an electric vehicle, see Fig. 15.2. The top event "vehicle does not start" can have causes in different parts of the vehicle. It is likely that different components

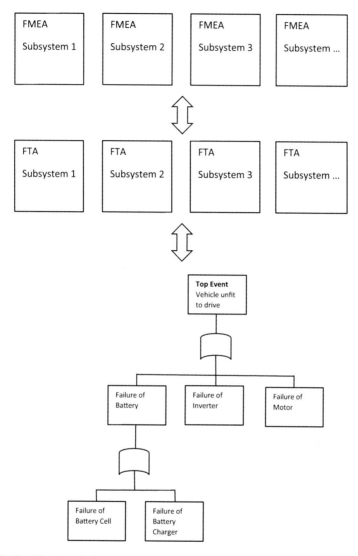

Fig. 15.2 Combination of subsystem FTAs to system FTA (Kanat 2014)

like the battery and the engine are analyzed individually and that the subtrees are later combined to a system analysis.

When combining subsystem fault trees to a system fault tree, it is important to consider interfaces between subsystems and dependencies of states. Note that Fig. 15.2 does not consider interfaces explicitly. Often replacement parts are available only at subsystem level. The final assessment needs to take the level of replacement into account.

After combining all subsystem fault trees, minimal cut sets of the whole system, and failure, causes for system failures can be determined.

Figure 15.2 shows a system, where no additional faults are expected due to the combination of subsystems. However, assume that the battery is close to a critical state (e.g., with respect to overall loading capacity) as well as the inverter (e.g., with respect to efficiency of the DC/AC conversion). The combination of both might lead to an overall system behavior that is critical in certain scenarios (e.g., long distances between loading stations or active safety systems that need high accelerations). Even more obvious are failures arising from interrupted interconnections of subsystems.

In Fig. 15.2, a simple example was chosen for illustration. The same method works when there are also other gates, for example, and gates.

Example Consider a system with three cordless phones (two docking stations and two identical, exchangeable receivers). Assume you defined the subsystems "docking station" and "receiver." For each of them, you set up a fault tree with the top event "subsystem is broken." To analyze your system and the top event "no phone calls possible," you have to analyze the interactions. For example, you have to consider failures where a damaged docking station damages one or both receivers. Hence, the probability that the system fails is not simply the sum of the probabilities that one of the components fails. It is also possible that you have an event in your subsystem fault tree for the receiver that says "does not receive electricity from station." In the system fault tree, this has to be linked with the subsystem fault tree of the docking station. You also have dependencies of events. For example, if there is no electricity, both docking stations will not work.

15.6   Fault Isolation Procedure

System-level fault trees can be used to create fault isolation procedures. The fault isolation procedure for an exemplary fault tree is shown in Fig. 15.3.

The shown example is simple because the fault tree only has OR gates. The fault isolation template can be realized similar to a questionnaire with statements like "If you answered with yes, please proceed to question 5." Especially in a digital version of the fault isolation template, this can be easily realized with filter questions. In a paper version, where one manually has to find the next question, this often leads to confusion.

It is more difficult to transfer a fault tree into a graphically supported fault isolation procedure if the fault tree comprises more different types of gates. One could, for example, add a horizontal level for AND gates and let the further direction in the template depend on the combinations of answers in a line. On paper, this is even more confusing than for OR gates and cannot be recommended. Priority And gates can only be described with text, without introducing special symbols.

FTA
Top Event Electrical motor cannot be actuated
Resolution according to lowest feasible level of diagnostics.

Example process to generate a symptom-based fault isolation procedure based on FTA	
Symptom: Electrical motor cannot be activated	
Motor cannot be activated (top event)	Three phase voltage operational (basic event) **Inverter does not start** (intermediate event) Inverter deactivated because of charging (intermediate event) ...
Inverter does not start	**Failure of start relay** (intermediate event) Connecting relay (basic event) Intermediate circuit voltage (intermediate event) Input fuse (basic event)
Start relay cannot be switched on	**Failure of start relay** (basic event) Steering signals (basic event)
Final diagnostic: Failure of start relay	

Fig. 15.3 Transformation of a fault tree into a fault isolation procedure (Kanat 2014)

When constructing the template, it has to be kept in mind that it is possible that several failures can occur simultaneously. Hence, the template must have filter questions where more than one answer is possible.

In Fig. 15.3, the light blue boxes show a minimal cut set of the fault tree. Transformed into the fault isolation procedure, the sheet can be used to help find the cause of the non-functioning inverter. The resulting cause in the example is the basic event belonging to the minimal cut set in the fault tree.

15.7 Questions

(1) What are the main mutual benefits when conducting an FMEA and an FTA jointly?
(2) Is it sufficient to conduct for subsystems FMEAs and to use FTA only on system level?
(3) How can you use an FTA to generate an efficient fault isolation procedure for the top event?

15.8   Answers

(1) See Fig. 15.1 and its discussion.
(2) See Fig. 15.2 and its discussion.
(3) Show how the stepwise assessment process of Fig. 15.3 can be derived by resolving for an FTA graphic top-down and stepwise (event layer by layer) the basic events by diagnostic approaches. This determines which basic events corresponding to faults need to be considered next. Note that the result may also be several diagnosed faults.

References

Automotive Word (2015): Bosch: DINA research project successfully completed – Consortium researches first integrated diagnostic system for electromobility. Available online at https://www.automotiveworld.com/news-releases/bosch-dina-research-project-successfully-completed-consortium-researches-first-integrated-diagnostic-system-electromobility/, updated on 12/22/2015, checked on 9/21/2020.
Barcin, Bülent; Freuer, Andreas; Kanat, Bülent; Richter, Andreas (2014): Wettbewerbsfähige Diagnose und Instandsetzung. In *ATZ Extra* 19 (11), pp. 14–19. https://doi.org/10.1365/s35778-014-1285-6.
David, P., V. Idasiak and F. Kratz (2008). Towards a better interaction between design and dependability analysis: FMEA derived from UML/SysML models. European Safety and Reliablity Conference (ESREL 2009) Safety, Reliability and Risk Analysis: Theory, Methods and Applications. S. Martorell, C. G. Soares and J. Barnett, CRC Press/Balkema, Taylor and Francis: 2259–2266.
IEC 61508 S+ (2010). Functional Safety of Electrical/Electronic/Programmable Electronic Safety-related Systems Ed. 2 Geneva, International Electrotechnical Commission.
Jain, Aishvarya Kumar; Satsrisakul, Yupak; Fehling-Kaschek, Mirjam; Häring, Ivo; van Rest, Jeroen (2020): Towards Simulation of Dynamic Risk-Based Border Crossing Checkpoints. In Piero Baraldi, Francesco Di Maio, Enrico Zio (Eds.): Proceedings of the 30th European Safety and Reliability Conference and the 15th Probabilistic Safety Assessment and Management Conference. ESREL2020 and PSAM15. European Safety and Reliability Aassociation (ESRA), International Association for Probabilistic Safety Assessment and Management (PSAM). Singapore: Research Publishing Services. Available online at https://www.rpsonline.com.sg/proceedings/esrel2020/pdf/4000.pdf, checked on 9/25/2020.
Kanat, B. (2014). FTA - Personal communication with S. Rathjen.
Kanat, B; Ebenhöch, S (2015): BMBF-Verbundprojekt DINA - Diagnose und Instandsetzung im Aftersales für Elektrofahrzeuge, Teilvorhaben: Zuverlässigkeitsvorhersagen von "High Voltage"-Systemen im Elektrofahrzeug für die Umsetzung von Diagnoseservices. Diagnose und Instandsetzung im Aftersales für Elektrofahrzeuge, Teilvorhaben: Zuverlässigkeitsvorhersagen von "High Voltage"-Systemen im Elektrofahrzeug für die Umsetzung von Diagnoseservices. With assistance of TIB-Technische Informationsbibliothek Universitätsbibliothek Hannover, Technische Informationsbibliothek (TIB). Edited by Fraunhofer EMI (Bericht E 38/2015). Available online at https://www.tib.eu/de/suchen/id/TIBKAT:881610984/, checked on 9/21/2020.
Larisch, M., A. Hänle, I. Häring and U. Siebold (2008a). Unterstützung des Nachweises funktionaler Sicherheit nach IEC 61508 durch SysML. Dipl. Inform. (FH), HTWG-Konstanz.
Larisch, M., A. Hänle, U. Siebold and I. Häring (2008b). SysML aided functional safety assessment. Safety Reliablity and Risk Analysis: Theory, Methods and Applications, European Safety and

Reliablity Conference (ESREL) 2008. S. Martorell, C. G. Soares and J. Barett. Valencia, Spanien, Taylor and Franzis Group, London. **2**: 1547–1554.

Larisch, Matthias; Siebold, Uli; Häring, Ivo (2009): Assessment of functional safety of fuzing systems. In: International system safety conference // 27th International System Safety Conference and Joint Weapons System Safety Conference 2009 (ISSC/JWSSC 2009). Huntsville, Alabama, USA, 3–7 August 2009. Huntsville, Alabama, USA: Curran.

Li, G. and B. Wang (2011). SysML Aided Safety Analysis for Safety-Critical Systems. Artificial Intelligence and Computational Intelligence.

Liu, Chi-Tang; Hwang, Sheue-Ling; Lin, I-K. (2013): Safety Analysis of Combined FMEA and FTA with Computer Software Assistance – Take Photovoltaic Plant for Example. In *IFAC Proceedings Volumes* 46 (9), pp. 2151–2155. https://doi.org/10.3182/20130619-3-ru-3018.00370.

Mhenni, F., N. Nguyen and J.-Y. Choley (2014). Automatic fault tree generation from SysML system models. Advanced Intelligent Mechatronics (AIM). Besacon: 715–720.

Peeters, J.F.W.; Basten, R.J.I.; Tinga, T. (2018): Improving failure analysis efficiency by combining FTA and FMEA in a recursive manner. In *Reliability Engineering & System Safety* 172, pp. 36–44. https://doi.org/10.1016/j.ress.2017.11.024.

Renger, P; Siebold, U; Kaufmann, R; Häring, I (2015): Semi-formal static and dynamic modeling and categorization of airport checkpoints. In Tomasz Nowakowski (Ed.): Safety and reliability. Methodology and applications; [ESREL 2014 Conference, held in Wrocław, Poland. ESREL. London: CRC Press.

Schoppe, C A; Häring, I; Siebold, U (2014): Semi-formal modeling of risk management process and application to chance management and monitoring. In R. D. J. M. Steenbergen (Ed.): Safety, reliability and risk analysis: beyond the horizon. Proceedings of the European Safety and Reliability Conference, Esrel 2013, Amsterdam, The Netherlands, 29 September–2 October 2013. Proceedings of the European Safety and Reliability Conference. Boca Raton, Fla.: CRC Press, pp. 1411–1418.

Schoppe, C; Zehetner, J; Finger, J; Siebold, U; Häring, I (2015): Risk assessment methods for improving urban security. In Tomasz Nowakowski (Ed.): Safety and reliability. Methodology and applications; [ESREL 2014 Conference, held in Wrocław, Poland. ESREL. London: CRC Press, pp. 701–708.

Shafiee, Mahmood; Enjema, Evenye; Kolios, Athanasios (2019): An Integrated FTA-FMEA Model for Risk Analysis of Engineering Systems: A Case Study of Subsea Blowout Preventers. In *Applied Sciences* 9 (6), p. 1192. https://doi.org/10.3390/app9061192.

Siebold, U. and I. Häring (2012). Semi-formal safety requirement specification using SysML state machine diagrams. ESREL. Helsinki, Finland.

Siebold, U.; Larisch, M.; Häring, I. (2010): Using SysML Diagrams for Safety Analysis with IEC 61508. In: Sensoren und Messsysteme. Nürnberg: VDE Verlag GmbH, pp. 737–741.

Siebold, Uli; Larisch, Matthias; Häring, Ivo (2009): SysML modeling of safety critical multi-technological system. In R. Bris, C. Guedes Soares, S. Martorell (Eds.): European Safety and Reliablity Conference (ESREL) 2009. Prague, Czech Republic.: Taylor and Franzis Group, London, pp. 1701–1706.

Chapter 16
Error Detecting and Correcting Codes

16.1 Overview

"Error-correcting coding is the art of adding redundancy efficiently so that most messages, if distorted, can be correctly decoded" (Pless 2011). The topic was first addressed in 1948 and has since then become a branch of electrical engineering. This can be addressed as method to control statistical hardware failures in case of memory bit flips or bit flips in messages in the context of functional safety, see Sect. 11.8.8

In Sect. 16.2, we introduce the parity bit. Section 16.3 shows how this simple concept is extended to the Hamming code, for example, Hamming(7,4). In Sect. 16.4, we introduce another error detection method, cyclic redundancy check (CRC) checksums. Based on this introduction we share the example by (Kaufman et al. 2012) in Sect. 16.5 who analyze different encoding and error-detecting schemes for a sample system. Finally, we state how error-detecting codes are discussed in the standard IEC 61508, see Sect. 16.6.

The main sources for this chapter are (Kaufman et al. 2012) and (Hamming 1950).

The results from the sample system in Sect. 16.4 confirm the theoretical thoughts from Sect. 16.3 that error correction becomes difficult for errors in more than one bit. The results also confirm that all one-bit errors can be corrected through Hamming(7,4), given that it is known that there is only one error.

16.2 Parity Bit

The simplest method of error detection is the parity bit. The parity bit counts the number of "1"s in a binary string. The parity bit can be "even" or "odd."

Example The number $13 = 8 + 4 + 0 + 1$ in binary string is 1101. The parity bit would be "odd" because there are three "1"s.

© The Author(s), under exclusive license to Springer Nature Singapore Pte Ltd. 2021
I. Häring, *Technical Safety, Reliability and Resilience*,
https://doi.org/10.1007/978-981-33-4272-9_16

There are several variations of parity bits, for example, considering parity bits of substrings.

16.3 Hamming Code

An example of perfect error correction using Hamming(7,4) reads.

Message: 1001
Encoder: 0,011,001 (fat: original message)
Channel with noise transmits message
Received message: 0,011,011 (underlined: single bit-flip)
Decoder realizes that two of the control bits, so called parity bits (non-fat numbers in encoded message) are wrong (inconsistent with the overall message and protocol) and constructs an uncorrupted message 0,011,001 (assuming a single bit-flip in this case)
Decoded message: 1001.

The simple parity bit can only be used to find out that there are an uneven number of bit errors without locating them in the string. The Hamming code improves this by using several parity bits.

We explain the idea behind the Hamming code for the Hamming(7,4) code based on (Hamming 1950).

For Hamming(7,4), we have seven bits c_i of which four are data bits d_j and three parity bits p_k, see Table 16.1. The reason for the order of the data and parity bits is the binary number of the i in c_i. For example, the first parity sum sums all c_i where the i has a 1 in the very right position ($3 = 011, 5 = 101, 7 = 111$), the second parity sum sums all c_i where the i has a 1 in the second position from the right ($3 = 011, 6 = 110, 7 = 111$), etc. This can be extended for more data bits (Hamming 1950).

For Hamming(7,4), the parity bits are computed as follows (Hamming 1950):

$$p_1 = d_1 + d_2 + d_4, \tag{16.1}$$

$$p_2 = d_1 + d_3 + d_4, \tag{16.2}$$

$$p_3 = d_2 + d_3 + d_4. \tag{16.3}$$

Table 16.1 Distribution of data and parity bits for Hamming(7,4)

c_1	c_2	c_3	c_4	c_5	c_6	c_7
p_1	p_2	d_1	p_3	d_2	d_3	d_4

Error in this bit	Shows in which parity bits
d_1	p_1, p_2
d_2	p_1, p_3
d_3	p_2, p_3
d_4	p_1, p_2, p_3
p_1	p_1
p_2	p_2
p_3	p_3

Table 16.2 One-bit errors and their consequences

Example If there is a bit flip in d_3, the parity bits p_2 and p_3 do no longer match the sums in (16.2) and (16.3). Hence, the bit flip can be noticed in those two parity bits.

Table 16.2 shows that, given that there are only one-bit errors, an error in one bit of the overall string can be uniquely identified in the parity bits. There is a unique pattern at the right-hand side of Table 16.2 that can be uniquely mapped to the left-hand side of Table 16.2. That is, if there are only one-bit errors, they cannot only be identified but also corrected.

A well-defined unique operational approach derived from Table 4.2 reads as follows. After the data string transmission, the program computes the parity bits according to the Hamming(7,4) scheme using (16.1) to (16.3). According to Table 4.2, and assuming a one-bit flip of any of the string bits:

- If p_1 and p_2 do not fit to the data (i.e., the d_j), d_1 is erroneous and has to be corrected (i.e., flipped), etc.
- If p_1, p_2, and p_3 do not fit to the data, d_4 is erroneous and has to be corrected.
- If p_1 does not fit to the data, do not correct any of the data bits, etc.

If there are up to two-bit errors, it can still be identified that there is an error but, in general, not which type and where it is. Hence, a correction is not possible.

Example If d_1 and d_2 are wrong, p_1 is correct (since the variables only take the values 0 and 1) and p_2 and p_3 are wrong. But p_2 and p_3 being wrong could also mean that only d_3 is incorrect. Hence, assuming one- and two-bit errors are occurring, identifying that p_2 and p_3 are incorrect does not tell which data bits are incorrect.

Even if there are only two-bit errors, it can still be identified in all cases that a two-bit error occurred. Only in most of the cases, it can be identified that a two-bit error occurred in the data bits! In none of the cases, it can be detected where they occurred and hence how to correct them!

Example When inspecting the first and last line of Table 16.3 it becomes obvious that it cannot be decided whether the data bits (d_1 and d_2) or the parity bits (p_2 and p_3) have been both corrupted.

Table 16.3 shows all combinations of two-bit errors and in which parity bits can be seen.

Table 16.3 Two-bit errors and their consequences, for example, Hamming(7,4)

Erroneous bits							Error shows in		
d_1	d_2	d_3	d_4	p_1	p_2	p_3	p_1	p_2	p_3
x	x							x	x
x		x					x		x
x			x						x
x				x				x	
x					x		x		
x						x	x	x	x
	x	x					x	x	
	x		x					x	
	x			x					x
	x				x		x	x	x
	x					x	x		
		x	x				x		
		x		x			x	x	x
		x			x				x
		x				x	x		
			x	x				x	x
			x		x		x		x
			x			x	x	x	
				x	x		x	x	
				x		x	x		x
					x	x		x	x

So far, we have discussed cases of one- or two-bit errors. We shortly summarize the situation for more than two-bit errors:

- Three-bit errors are not always identified: If there are bit flips in d_1, d_4, and p_3, the errors even each other out in the parity sums and the errors are not noticed.
- The same holds for four-bit errors, for example, bit flips of d_4, p_1, p_2, and p_3 are not noticed.
- The situation of five-bit errors is symmetric to two-bit errors. They are always detected (not how many but that there are errors) but cannot always be corrected.
- The situation of six-bit errors is symmetric to one-bit error. The six-bit errors are always detected and can be corrected, given that one knows that there are six-bit errors.
- Seven-bit errors are not detected, all three equations still hold.

Remark (Hamming 1950): An extended code that corrects one-bit errors and detects two-bit errors can be constructed by adding another parity number that adds all c_i. The additional number can tell whether there is an even or odd number of bit flips.

Exemplary codes can be found online. The code used for the Hamming error-correction method in Sect. 16.4 is from (Morelos-Zaragoza 2006, 2009).

16.4 CRC Checksums

Another possibility of error detection is so-called CRC (cyclic redundancy check) checksums where data is encoded with a fixed binary polynomial which has to be known to the sender and receiver of the data. It is best understood if one regards an example.

Example We fix the CRC polynomial $x^4 + x^3 + 1 = 1x^4 + 1x^3 + 0x^2 + 0x + 1$, that is, 11001 in binary code.

As data that we want to send we take 10010011.

The first step is to attach as many zeros to the data as the degree of the polynomial, in our case, four: 10010011 0000.

Next, we divide this long binary number by the binary code of the CRC polynomial:

$$
\begin{array}{l}
100100110000/11001 = 11101001 \\
\underline{11001} \\
010110 \\
\quad\underline{11001} \\
\quad011111 \\
\quad\ \underline{11001} \\
\quad\ 001101 \\
\quad\quad\underline{00000} \\
\quad\quad0011010 \\
\quad\quad\ \underline{11001} \\
\quad\quad\ 000110 \\
\quad\quad\quad\underline{00000} \\
\quad\quad\quad0001100 \\
\quad\quad\quad\ \underline{00000} \\
\quad\quad\quad\ 00011000 \\
\quad\quad\quad\quad\underline{11001} \\
\quad\quad\quad\quad00001
\end{array}
$$

The important part is the remainder of the division. It is added to the data instead of the four zeros: 10010011 ~~0000~~ **0001.**

So, the binary code we transmit is 100100110001.

By construction, this number is divisible by 11001 without remainder. So, the received message can be divided by 11001 and if there is no remainder, the message is very likely to be correct.

Four cases can occur as follows:

- The transmitted message is correct and the remainder of the division by the CRC polynomial is 0.
- The transmitted message is incorrect but the remainder of the division by the CRC polynomial is not 0 (very unlikely). For one-bit errors, this case is not possible if the CRC polynomial's binary string contains at least two "1"s.
- The transmitted data is incorrect and the remainder of the division is not 0.
- Only the transmitted remainder is incorrect (the data itself is correct) and the remainder of the division is not 0. (This cannot be distinguished from the previous case.)

We illustrate the four cases in a short example:

Example CRC polynomial $x + 1$, that is, 11; data 100.

$$10000/101 = 101$$
$$\underline{101}$$
$$00100$$
$$\underline{101}$$
$$001$$

Case 1: Everything is correctly transmitted and the remainder is 0

$$10001/101 = 101$$
$$\underline{101}$$
$$00100$$
$$\underline{101}$$
$$000$$

Case 2: There were transmission mistakes (11011) but the remainder is still 0.

$$11011/101 = 111$$
$$\underline{101}$$
$$0111$$
$$\underline{101}$$
$$0101$$
$$\underline{101}$$
$$000$$

Case 3: Transmission mistake in data (10101) and the remainder is not 0

$$10101/101 = 100$$
$$\underline{101}$$
$$00001$$

Case 4: Transmission mistake in attachment (10011) and the remainder is not 0

$$10011/101 = 101$$
$$\underline{101}$$
$$00111$$
$$\underline{101}$$
$$010$$

16.5 Assessment of Bit Error-Detecting and Error-Correcting Codes for a Sample System

In this section, we show an example of applying error-detecting and error-correcting codes to a sample system and an analysis thereof.

16.5.1 The Sample Problem

Kaufman et al. (2012) analyze the following procedure of transmitting a time value from a programming tool to a microprocessor.

- Programming tool with data transmission interface:

    Step 1: Enter time value t_{prog} [value between 0.0 and 999.9 s in increments of 0.1 s].
    Step 2: Store as floating point.
    Step 3: Encoding 1 of 2: Convert to packed BCD (Binary Coded Decimal).
    Step 4: Further encoding 2 of 2 (as needed): Hamming, 2oo3, XOR;

- Transmission of encoded data to EEPROM (Electrically Erasable Programmable Read-Only Memory);
- Transmission of encoded data to microprocessor.
- In microprocessor:

    Step 5: Decoding (in the following cases): Hamming, XOR.
    Step 6: Convert packed BCD to float.
    Step 7: Received time value t'_{prog}.

The time value t_{prog} is a "two-byte packed binary coded decimal (BCD)" (Kaufman et al. 2012), see Fig. 16.1.

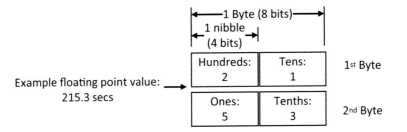

Fig. 16.1 Packed BCD representation of four-digit decimal value for an example value of 215.3 s (Kaufman et al. 2012)

16.5.2 Assumptions

To analyze bit error-detecting codes, Kaufman et al. (2012) "make the following assumptions:

- Each bit has the same probability of flipping to the incorrect state.
- Only bit errors are considered, i.e., no bit shifts.
- Bit errors occur statistically independent of each other, common-cause errors are not considered, e.g., any single radiation event only affects one bit. There is no spatial local clustering of damage events.
- Each bit can only have a single bit error, i.e. a double bit error must involve two distinct bits and cannot involve a single bit which first changes bit state and then changes bit state again.
- All initial time values within the available number range will be considered [without assuming any possibly existing (set of) problem specific intervals] [...]. We test 10,000 values ranging from 0.0 to 999.9 in 0.1 increments.
- We do not discern between soft (transient or intermittent) and hard (permanent) bit errors."

16.5.3 The Simulation Program and Running Time

For the analysis, they wrote a simulation program with different error-detecting codes:

"A simulation program was written and executed in C++ using Visual Studio Premium 2010 on a PC outfitted with an AMD Athlon 64 × 2 processor. We implemented the Hamming(7,4) code, as published in (Morelos-Zaragoza 2006). The other Hamming methods tested were implemented by making minor adjustments to the original Hamming(7,4) code. We wrote the remaining methods BCD, XOR and BCD two out of three (2oo3) ourselves. [Fig. 16.2] shows the sequence of decisions of the program" (Kaufman et al. 2012).

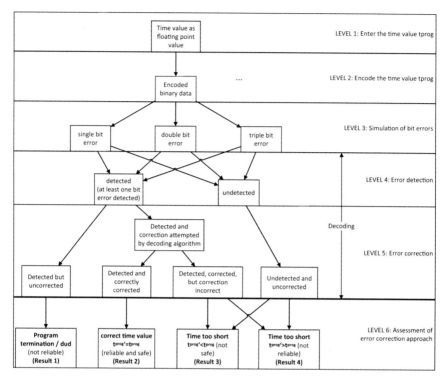

Fig. 16.2 Decision chart for error simulation program. *Source* Modified from (Kaufman et al. 2012)

In Table 16.4, we want to give examples for the four cases on Level 5 for the Hamming(7,4) code with an additional parity bit

Table 16.4 Examples of how a program for Hamming(7,4) with additional bit reacts

Case	Errors in	Shows in	Consequences
Detected but uncorrected	d_1, d_2	p_2, p_3	Because p_4 is correct, it is known that it is a two-bit error but could also be in d_4, p_1
Detected and correctly corrected	d_1	p_2, p_3, p_4	Because p_4 is incorrect, it is an odd number of bit errors. If the program assumes a one-bit error, it is correctly corrected
Detected, corrected, but correction incorrect	d_2, d_3, d_4	p_2, p_3, p_4	Program thinks it is in the case above and corrects d_1
Undetected and uncorrected	d_1, d_2, d_3, p_4	-	Program does not recognize error

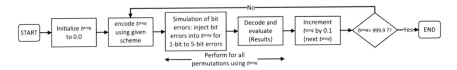

Fig. 16.3 Flow diagram for simulation program: variable t_{prog} (Kaufman et al. 2012)

$$p_4 = d_1 + d_2 + d_3 + d_4 + p_1 + p_2 + p_3 \qquad (16.4)$$

for a program that assumes as little errors as possible, that is, it believes that one-bit errors are the most likely to occur.

Figure 16.3 shows how the simulation program works. It starts with $t_{prog} = 0$. Then this fixed time value is encoded, bit errors are simulated, the value is decoded and it is evaluated whether the error-detecting code found the simulated bit errors (compare Fig. 16.2). Then the same is done with the next time value $t_{prog} = 0.1$ etc., up to $t_{prog} = 999.9$.

Kaufman et al. (2012) also considered how much memory was needed for the encoded data. For this, they distinguish between information bits and redundancy bits.

Example We go back to the example from Sect. 16.2. We had the binary string 1101 and one parity bit. In this case, the total number of bits n is 5, the number of information bits m is 4, and the number of redundancy bits k is 1.

In Fig. 16.4, the ration of information bits to the total number of bits m/n is used to compare the different encoding schemes used. Here $m = 16$, compare Fig. 16.1. Kaufman et al. (2012) "used the notation Hamming(n,m) for the single error correcting (SEC) Hamming code. Here the number of bits in the Hamming word corresponds to the total number of bits n, and m is the number of information bits. For the single error correcting double error detecting (SEC-DED) code, we use

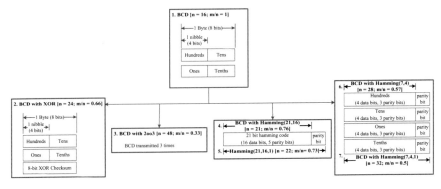

Fig. 16.4 Error-detection schemes to be tested and their corresponding amounts of allocated memory used (Kaufman et al. 2012)

the notation Hamming(*number of bits in Hamming word, number of information bits, additional parity bit*)."

Kaufman et al. (2012) distinguish between one-, two-, three-, four-, and five-bit errors. For each encoding method and each number of bit errors, they compute the number of possible failure combinations with the binomial coefficient.

Example For BCD with Hamming(21,16), there are 21 bits where the bit errors can occur. That makes for a fixed time value,

$$\binom{21}{1} = 21 \text{ combinations with one-bit error,}$$

$$\binom{21}{2} = 210 \text{ combinations with two-bit errors,}$$

$$\binom{21}{3} = 1330 \text{ combinations with three-bit errors,}$$

$$\binom{21}{4} = 5985 \text{ combinations with four-bit errors, and}$$

$$\binom{21}{5} = 20348 \text{ combinations with five-bit errors.}$$

There are 10,000 time values (000.0 to 999.9 in 0.1 steps). So, for example, the total number of combinations with three-bit errors is 1,330,000.

16.5.4 Results

Figures 16.5, 16.6, 16.7, 16.8 and 16.9 show which encoding schemes are most efficient for the sample system and one- to five-bit errors.

"The simulation program [...] only considered cases with bit errors. In reality, bit errors will only occur with a certain probability. As such, the probability of an error actually occurring needs to be overlaid with the above simulation. Here we explore the conditions under which we believe it is acceptable to use bit correcting techniques. Keeping in mind our assumption from [Sect. 16.5.2] that bit errors occur statistically independently, we let f be the frequency of a single bit error, e.g., in units of h^{-1}. In time interval T we claim that Tf is approximately the probability of a single bit error assuming $Tf < < 1$. Let r_i^{cc} be the ratio of the correctly corrected (green shading in [Figs. 16.5 to 16.9]) and r_i^{ec} of erroneous corrected (see [red] and blue shading) of the i injected bit errors in the n-bit data, and n_{bit} the maximum number of corrected bits. Then

$$P^{ec} = \sum_{i=1}^{n_{bit}} \binom{n}{i} r_i^{ec} (T f)^i (1 - T f)^{n-i} \tag{16.5}$$

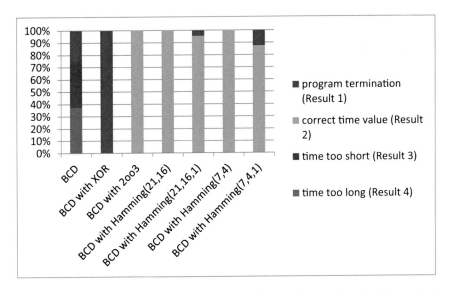

Fig. 16.5 Results of the bit error simulation program for one-bit errors for all coding methods tested (Kaufman et al. 2012)

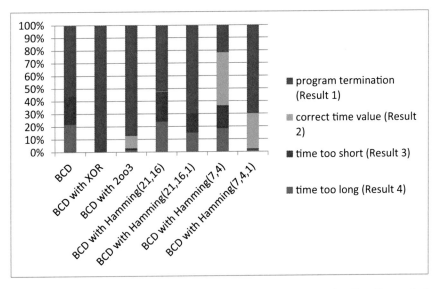

Fig. 16.6 Results of the bit error simulation program for two-bit errors for all coding methods tested (Kaufman et al. 2012)

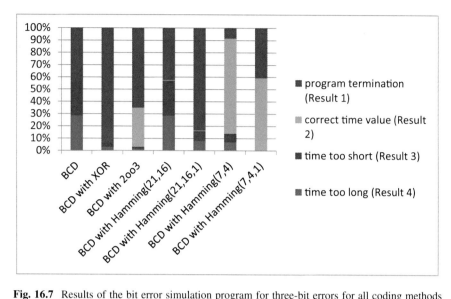

Fig. 16.7 Results of the bit error simulation program for three-bit errors for all coding methods tested (Kaufman et al. 2012)

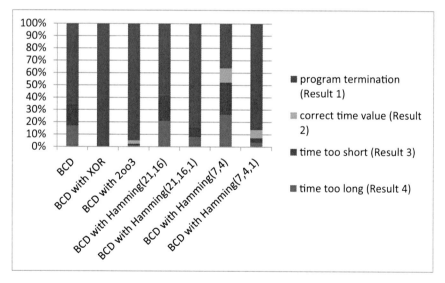

Fig. 16.8 Results of the bit error simulation program for four-bit errors for all coding methods tested (Kaufman et al. 2012)

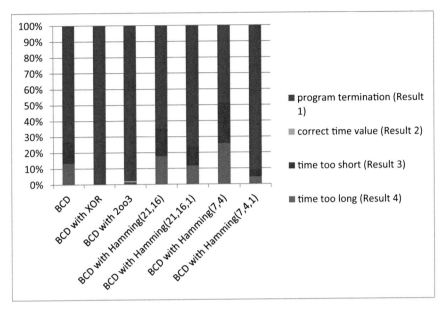

Fig. 16.9 Results of the bit error simulation program for five-bit errors for all coding methods tested (Kaufman et al. 2012)

is the probability that bit errors were erroneously corrected and the same expression for P^{cc} that they were correctly corrected. For a given [safety–critical] system, one has to show that the increase of the probability of reliability by P^{cc} is not offset by an intolerable increase of the probability […] P^{ec} due to erroneous correction" (Kaufman et al. 2012).

The results of Figs. 16.5 to 16.9 do not distinguish between not detected and detected but faultly corrected errors, compare Fig. 16.2. However, the percentage of detected errors is at least as big as the green and violet columns of Figs. 16.5 to 16.9 together. For Hamming(7,4), this percentage is only around 50–60% for three or more bit errors.

16.6   Error-Detecting and Error-Correcting Codes in the Standard IEC 61508

Error-detecting and error-correcting codes are also described in the IEC 61508:

"Aim: To detect and correct errors in sensitive information.

Description: For an information of n bits, a coded block of k bits is generated which enables r errors to be detected and corrected. Two example types are Hamming codes and polynomial codes.

It should be noted that in safety-related systems it will normally be necessary to discard faulty data rather than try to correct it, since only a predetermined fraction of errors may be corrected properly." (IEC 61508 2010).

The standard IEC 61508 Ed. 2 suggests the book by Huffman and Pless (2003).

In part 3 of the standard, Table A.2, error-detecting codes are recommended for SIL 1 to 3 and highly recommended for SIL 4.

16.7   Questions

(1)   Which error-detection and error-correction codes do you know? Explain their main ideas!
(2)   What has to be considered in case of use of error detection and correction in safety–critical applications, also according to IEC 61508?
(3)   Give an example that a parity checksum does not detect a multiple bit flip in a nibble!
(4)   Give an example that 2oo3 (2 out of 3) does not detect a two-bit flips in a nibble!
(5)   How can one systematically assess the detection and correction capability of Hamming codes?

16.8   Answers

(1)   See Sects. 16.2, 16.3, and 16.4 and respective explanations. See also the MooN redundancy introduced in Sect. 16.5.1.
(2)   Since automatic self-correcting codes are not recommended by IEC 61508, it has to be shown that correction is better than non-correction considering all possible cases, i.e., also multiple bit flips of higher order as well as the given operational context. See also the argumentation for the sample case in Sect. 16.5.
(3)   Nibble plus parity bit example: 11000. Example after bit flip: 000000.
(4)   Nibble example: 1100. Redundant data after double bit flip: 1000, 1000, 1100. Hence, 2oo3 will select 1000.
(5)   Options include a complete assessment of all combinations for expected parameter values, combinatorial analysis, or use a weighted Monte Carlo approach with focus on low-order number of bit flips and bit shifts.

References

Hamming, R. W. (1950). "Error Detecting and Error Correcting Codes." The Bell System Technical Journal **29**(2): 147–160.
Huffman, W. C. and V. Pless (2003). Fundamentals of Error-Correcting Codes, Cambridge University Press.

IEC 61508 (2010): Functional Safety of Electrical/Electronic/Programmable Electronic Safety-related Systems Edition 2.0. Geneva: International Electrotechnical Commission.

Kaufman, J. E., S. Meier and I. Häring (2012). Assessment of bit error detecting and correcting codes ESREL-conference 2012.

Morelos-Zaragoza, R. H. (2006). The Art of Error Correcting Coding, John Wiley & Sons.

Morelos-Zaragoza, R. H. (2009). "the-art-of-ecc.com." Retrieved 2014-11-04, from http://the-art-of-ecc.com/.

Pless, V. (2011). Introduction to the Theory of Error-Correcting Codes, John Wiley & Sons.

Bibliography

AS/NZS ISO 31000:2009: Risk management - Principles and guidelines.

Elamkulam, J., Z. Glazberg, I. Rabinovitz and G. Kowlali (2006). Detecting Design Flaws in UML State Charts for Embedded Software. Second International Haifa Verification Conference. Haifa, Israel.

Günther, H.-J. (2007). "Anwendung Fehler - Möglichkeits- und Einflussanalyse am Anwendungsbeispiel." Retrieved 2013-03-06, from http://www.mb.hs-wismar.de/~guenther/lehre/Qualitaetssicherung/QFD/Beispiel%20QFD_FMEA-Fahrrad.pdf.

Hardy, T. L. (2005). Integrating Launch Vehicle Safety and Reliability Analyses. 23rd International System Safety Conference.

Häring, I., S. Rathjen, S. Slotosch and S. Fényes. (2014). "Sicherheitssystemtechnik." Retrieved 2015-01-09, from http://www.offenehochschule.uni-freiburg.de/sicherheitssystemtechnik.

IEC 61508 Series, 2010: Functional safety of electrical/electronic/programmable electronic safety-related systems. Available online at http://www.iec.ch/functionalsafety/standards/page2.htm, checked on 12/27/2017.

IEC 61508, 2010: IEC 61508 - Functional safety of electrical/electronic/programmable electronic safety-related systems. Available online at https://www.iec.ch/functionalsafety/standards/page2.htm, checked on 9/12/2020.

Kaufman, J E; Meier, S; Häring, I (2012): Assessment of bit error detecting and correcting codes for safety-critical embedded fuzing systems. In: 11th International Probabilistic Safety Assessment and Management Conference and the Annual European Safety and Reliability Conference 2012. (PSAM11 ESREL 2012); Helsinki, Finland, 25–29 June 2012. International Association for Probabilistic Safety Assessment and Management; European Safety and Reliability Association; International Probabilistic Safety Assessment and Management Conference; PSAM; Annual European Safety and Reliability Conference; ESREL. Red Hook, NY: Curran, pp. 1869–1878.

Meier, Stefan (2011): Entwicklung eines eingebetteten Systems für eine Sensorauswertung gemäß IEC 61508. Bachelor Thesis. Duale Hochschule Baden-Württemberg, Lörrach. Fraunhofer EMI.

Moret, B. M. E.; Thomason, M. G. (1984): Boolean Difference Techniques for Time-Sequence and Common-Cause Analysis of Fault-Trees. In *IEEE Trans. Rel.* R-33 (5), pp. 399–405. https://doi.org/10.1109/tr.1984.5221879.

Rouse, M. (2007). "Spiral model (spiral lifecycle model)." Retrieved 2013-12-10, from http://searchsoftwarequality.techtarget.com/definition/spiral-model.

Sudheendran, Vivek (2020): SysML supported functional safety and cybersecurity assessment for automotive MBSE. Master Thesis. University of Freiburg, Freiburg. IMTEK.

Wikipedia. (2011). "V-Model." Retrieved 2012-03-22, from http://en.wikipedia.org/wiki/V-Modell.

© The Editor(s) (if applicable) and The Author(s), under exclusive license to Springer Nature Singapore Pte Ltd. 2021
I. Häring, *Technical Safety, Reliability and Resilience*,
https://doi.org/10.1007/978-981-33-4272-9

Wikipedia. (2012). "Dependency (UML)." Retrieved 2012-05-08, from http://en.wikipedia.org/wiki/Dependency_(UML).

Wikipedia. (2012). "Spiral model." Retrieved 2012-03-22, from http://en.wikipedia.org/wiki/Spiral_model.

Wikipedia. (2012). "Wasserfallmodell." Retrieved 2012-03-22, from http://de.wikipedia.org/wiki/Wasserfallmodell.

Wikipedia. (2012). "Waterfall model." Retrieved 2012-03-22, from http://en.wikipedia.org/wiki/Waterfall_model.

Wikipedia (2012). Bathtub curve.

Wikipedia (2012). Hazardrate.

Zheng, W. and G. Bundell (2007). A UML-based methodology for software component testing. International MultiConference of Engineers and Computer Scientists. Kowloon, China, Centre for Intelligent Information Processing Systems.

Zio, E., M. Librizzi and G. Sansavini (2006). "Determining the minimal cut sets and Fussell-Vesely importance measures in binary network systems by simulation." Safety and Reliability Managing Risk.

Index

© The Editor(s) (if applicable) and The Author(s), under exclusive license
to Springer Nature Singapore Pte Ltd. 2021
I. Häring, *Technical Safety, Reliability and Resilience*,
https://doi.org/10.1007/978-981-33-4272-9